THE STUDY OF THE HISTORY OF SCIENCE

BEING THE SUBSTANCE OF THE INAUGURAL LECTURE OF THE SEMINARY ON THE HISTORY OF SCIENCE IN HARVARD UNIVERSITY DELIVERED ON OCTOBER 4, 1935

LONDON : HUMPHREY MILFORD
OXFORD UNIVERSITY PRESS

THE STUDY OF THE HISTORY OF SCIENCE

BY

GEORGE SARTON
S.D., A.C.I., L.H.D., LL.D.

CAMBRIDGE · MASSACHUSETTS
HARVARD UNIVERSITY PRESS
1936

PRINTED AT THE HARVARD UNIVERSITY PRESS
CAMBRIDGE, MASS., U.S.A.

CONTENTS

THE STUDY OF THE HISTORY OF SCIENCE

THE STUDY OF THE HISTORY OF SCIENCE

THE publication of university lectures may not, in general, be desirable, because it is better that they be kept in as fluid a state as possible, but an exception should be made for inaugural lectures. Indeed, each of these lectures is, or should be, a milestone, and the examination of several of them relating to a single discipline enables us to measure the progress of that subject in a manner doubly pleasant, because concrete and informal. As it is my privilege to inaugurate, not only this seminary, but also the scientific study of the history of science in Harvard University, I feel it to be a part of my duty to explain my task as I understand it. I hope that in so doing I am beginning a new tradition, and that my successors will take pains to explain the same task, as they understand it, when it shall be their privilege in the course of time to assume it.

One might assert that there is nothing new in the history of science, because it is simply the application of the well known methods of history to the well known facts of science. This is specious. The elements of our discipline may be as old as you please, their combination is relatively new. It is true some historical efforts were made as early as the fourth century B.C. by Aristotle's pupil, Eudemos of Rhodes, and these efforts were continued by other Greeks; furthermore there has come down to us from the Middle Ages and modern times a whole series of writings in Arabic, Latin, and other languages, which might be catalogued under the heading 'history of science.' Nevertheless, the first scholar to conceive that subject as an independent discipline and to realize its importance was the French philosopher Auguste Comte (1798–1857), and the scholar who deserves perhaps more than any other to be called the father of our studies,

Paul Tannery (1843–1904), could write, as late as 1904, "To-day this history [i. e., the history of science] is nothing but an individual conception." [1] What Tannery meant was that that history was not yet represented by an array of noble books constituting the tradition of a definite kind of scholarship different from all others. One wonders whether he was acquainted with Whewell's work. The chances are he was not.[2] In any case Whewell's conception of the history of science was still primitive and somewhat narrow, and his realization of it imperfect. Postponing the discussion of his priority, it is clear that the history of science is a relatively new discipline. Whether it be fifty years old, or a hundred, it is still so young that I should not hesitate to repeat Tannery's statement. It is not yet crystallized in the sense that the 'history of England' is, or even the 'history of religion.' Not to speak of an 'array' of books, there is not yet a single one which is so good that one can refer to it to the exclusion of others, not yet a single one which is truly worthy of the subject. We are still turning around it, approaching and investing it from every side; we do not yet master it.

* * *

In fact, most scholars misunderstand its true nature. This is not surprising, in view of its dual origin. Whom shall we consider the best prepared and most expert, him who understands the subject or him who knows the methods? In the *history* of *science* shall we emphasize the first word or the second? It is clear also that our definition of the new discipline will be colored by our conception, or miscon-

[1] Paul Tannery, "De l'histoire générale des sciences," *Revue de synthèse historiques*, vol. 8 (1904), p. 7.

[2] William Whewell (1794–1866). His *History of the Inductive Sciences from the Earliest to the Present Times* (1837, 3 vols.) was as little known in the French world as it was well known in the English world. It was soon translated into German (1840–41) but not into French.

ception, of historiography on the one hand and of science on the other, and even by our philosophical beliefs or prejudices.

Before proceeding further, it may be useful to repeat a definition and a theorem which I have published in various forms in earlier writings since 1913.

Definition. Science is systematized positive knowledge, or what has been taken as such at different ages and in different places.

Theorem. The acquisition and systematization of positive knowledge are the only human activities which are truly cumulative and progressive.

Corollary. The history of science is the only history which can illustrate the progress of mankind. In fact, progress has no definite and unquestionable meaning in other fields than the field of science.

To be sure, we should not be dazzled by the shibboleth 'progress,' for there are other features of human life which are at least as precious as scientific activities, though they are unprogressive: charity and the love of beauty, for example. But if we wish to explain the progress of mankind, then we must focus our attention on the development of science and its applications.

Moreover, we shall not be able to understand our own science of to-day (I do not say to use it, but to understand it) if we do not succeed in penetrating its genesis and evolution. Knowledge is not something dead and static, but something fluid, alive, and moving. The latest results are like the new fruits of a tree; the fruits serve our immediate practical purposes, but for all that it is the tree that matters. The scientist of philosophic mind is not interested so much in the latest results of science as he is in its eternal tendencies, in the living and exuberant and immortal tree. The fruits of to-day may be tempting enough, but they are not more precious to his way of thinking than those of yesterday or to-morrow.

One might thus expect the more intelligent scientists to appreciate the historical point of view, but most of them are so busy, so anxious to be always at the very frontier of expanding knowledge, that they have neither inclination nor time to look backward. Even when they are historically-minded, which is seldom the case, they are so keenly aware of their purely scientific difficulties that they are prone by way of contrast to minimize, or to overlook entirely, the historical ones. Having finally reached a summit whence they could dominate the past, either they consider retrospection an unwarranted indulgence and dereliction of duty, or they look backward so clumsily that they cannot see anything with precision. You may remember Littré's earnest words in the first volume of his magnificent edition of Hippocrates:

> Quand on s'est pénétré de la science contemporaine, alors il est temps de se tourner vers la science passée. Rien ne fortifie plus le jugement que cette comparison. L'impartialité de l'esprit s'y développe; l'incertitude des systèmes s'y manifeste; l'autorité des faits s'y confirme, et l'on découvre, dans l'ensemble, un enchaînement philosophique qui est en soi une leçon. En d'autres termes, on apprend à connaître, à comprendre, à juger.[1]

"When one has imbued one's self with the scientific knowledge of to-day, *then it is time* to turn toward the science of the past" . . . but the poor devils have no time to study the past, nor have they the faintest idea of how to study it. As the scientific training is enormously more difficult, and requires the unintermitting effort of years and years, indeed, of a lifetime, they are prone to think that it is sufficient. Alas! Indispensable as it is, it is not enough. The historical training is, or may be, much easier than the scientific one, but it is equally necessary.

* * *

The obvious difference between the scientific preparation and the historical one is that the former is not only much

[1] Emile Littré, *Oeuvres complètes d'Hippocrate* (Paris, 1839), vol. 1, p. 477.

longer but more systematic; it must be carried on in a certain order. One cannot study analysis before algebra, nor physiology before chemistry and physics. On the other hand, it is possible, though perhaps unadvisable, to study history in almost any order, and the majority of scholars have obtained their historical knowledge in the most haphazard way. One may be an expert in American history and know nothing whatever of the Sumerians or the Hittites. Some time ago I was obliged to examine the work of a Transylvanian chemist who flourished at the end of the eighteenth century. To appreciate his work it was necessary to consider on the one hand the historical *milieu*, and on the other the contemporary chemical traditions. I knew nothing of the history of Transylvania at that time, but it did not cost me much trouble to obtain the information which I needed, or, with the aid of my knowledge of comparable conditions elsewhere, to understand it. Happily I was well acquainted with chemistry and the chemical knowledge of that period in Western Europe and with the so-called 'chemical revolution,' subjects in which it would have been impossible for me to remedy my ignorance so readily. I should have been obliged to study chemistry and the complete history of chemistry down to that time!

However, the most fundamental difference between historical knowledge and scientific knowledge is revealed by the way they grow. Historical knowledge grows slowly and precariously; precariously, because of the constant recurrence of discredited errors; slowly, because of increasing difficulty in obtaining new material. Though our knowledge tends toward completeness, it is asymptotic, and never reaches the goal. For example, consider our knowledge of ancient Greece. It continues to improve, to be sure, but with smaller and smaller increments of truth. It will not be more difficult, it will probably be simpler, to study it a few centuries hence than now. On the contrary, any branch of science may be completely revolutionized at any time by a discovery necessitating a radically new approach to the subject. Chemistry to-day is essentially different from chemistry in the eighteenth century. The fundamental notions are different, the methods are different, the scope is incredibly larger, and the contents infinitely more varied. We may safely assume that the chemistry of the twenty-fifth century will be as unlike that of the present as that, in turn, is unlike that of the fifteenth century. On the other hand, in the twenty-fifth century or in the thirtieth it will take about the same pains and time as to-day to study Latin grammar, Greek literature, or the history of the eighteenth century. The literary subjects, we may say, tend to be closed subjects, whose expansion after a certain point is so slow as to be imperceptible. In strong contrast with them, the growth of scientific subjects is unpredictable, luxuriant, and sometimes explosive in its intensity and destructiveness.

The scientist, then, can never relax in his efforts and enjoy himself, like a genial and sensible grammarian, but must be prepared to learn new things every day, and, what is worse, unlearn others with which he has grown intimate, and change the tenets of a life time on fundamental points. No wonder that such a harassed individual is generally unwilling to contemplate the past, or, should he have any velleities to do so, unable to do it well. He innocently believes, it may be, that he knows how to do it. Historical work, he seems to think, consists in taking a few old books and copying from them this and that. He may be well trained and fastidious in his own exacting technique, yet not realize that the technique of establishing the truth, or the maximum probability, of past events has its own complicated rules and methods. Historical work, as he conceives it in his candor, is exceedingly easy; almost all that is needed, he thinks, is to know how to read and write, and he despises it accordingly. He does not realize that he is merely despising his perverted image of it. The historian whom he scorns and ridicules is nobody but himself.

* * *

The difference between the historical and the scientific points of view has been amusingly illustrated by Henri Poincaré as follows: [1]

> Carlyle has written somewhere something like this:
>
> "Nothing but facts are of importance. John Lackland passed by here. Here is something that is admirable. Here is a reality for which I would give all the theories in the world." Carlyle was a countryman of Bacon . . ., but Bacon would not have said that. That is the language of the historian. The physicist would say rather: "John Lackland passed by here; I don't care, for he will not pass this way again."

Physical sciences deal with the 'laws of nature,' with the repetition of facts under given circumstances, not only in the past but also in the future; history deals with isolated facts of the past, facts which cannot be repeated and hence cannot be thoroughly verified. At first view it seems impossible to bridge that abyss. And yet the difference is perhaps quantitative rather than qualitative. For, on the one hand, historical facts are more or less repeated. When a tyrannical rule is introduced into a country, one or more of certain well known series of events are bound to happen in consequence of it. The repetition is not complete and detailed, as in the case of physical or chemical facts, yet there is a repetition of patterns which deserves to be taken into account. The trouble with the Carlyle-Poincaré example is that it is too particular; it would be too particular even for the physicist. John Lackland will never come again; but there may be others like him, and patterns of a definite kind will entail a succession of other definite patterns. On the other hand, on account of the infinite complexity of causes and of the dissipation of energy, physical events never repeat themselves exactly. The planets do not follow twice the same trajectories.

The old saying of Heraclitus is truer than ever: Πάντα ῥεῖ,

[1] *La Science et l'Hypothèse*, p. 168.

everything flows. The physical world is less regular and the social world more regular than one generally admits, and thus the two are not so widely apart as we imagine.

As opposed to the more exact mathematical sciences, the historical 'sciences' seem to usurp their name, but it is not fair to compare the extremes of a series. A comparison with the natural sciences is more adequate. The historian of science, to return to him, is a collector of scientific ideas in the same way that the entomologist is a collector of insects, the 'collection' in both cases being only the first step along the road to knowledge. The point is that both will use similar methods to make sure that the items of their collections are as unequivocally and completely determined as possible, and when the facts are duly established they must needs use similar methods to draw their conclusions and to build up progressively a system of knowledge. The comparison of the historian with the naturalist might be pursued retrospectively at different stages of their growth. There was a time of innocence when their methods were equally immature and inconclusive; both have learned gradually, very gradually, to make the most of the available evidence, the most but not more, and even to measure to some extent their approximation to the truth. Under the healthful influence of geological and prehistoric research, some historians have now become full brothers to naturalists.

The prehistorians and other archaeologists have built a solid bridge between history and science, and we, historians of science, are now proceeding to build another one, even more substantial, and thus to help span the chasm which is cutting our culture asunder and threatening to destroy it. The scientific spirit is as much improved and purified by the admixture of historical considerations as is humanism itself by the introduction of scientific methods.

* * *

The main point to emphasize — and if this is properly understood all the rest follows without difficulty — is that *accuracy is as fundamental in the historical field as in the scientific one, and that it has the same meaning in both fields.* Experienced historians may find it strange that I should trouble to explain what to them is obvious, but it is necessary to do so in order to eradicate the prejudices of scientists against us, and enable them to come and meet the historians of science half way, instead of throwing spokes into our wheels.

Let us suppose that a physicist has to measure the length or distance *AB*. He may state that *AB* is 3 m. long, or 300 cm., or 3000 mm. These are three different statements, for they imply different degrees of accuracy: in the last case, for example, that the length is correct within a millimetre (2999 mm. < *AB* < 3001 mm.). The degree of accuracy obtainable or desirable varies with the circumstances, but one must be accurate within the limits which are appropriate and which are suggested by the choice of units or the number of decimals.

The situation is exactly the same with regard to dates. If you state that a certain event happened on October 20, 1495 (Gregorian), you must be reasonably sure that it is the 20th, not the 19th or the 21st. A scientist, however meticulous he may be in his own field, will shrug his shoulders and grumble: "What do I care whether it is the 20th or the 25th. . . ." Very well. It may indeed be of no importance, but then why state the day? Why not say "October, 1495," or "1495," or "toward the end of the fifteenth century"? The last statement might be the best one. To say "Oct. 20, 1495," when you are not sure of the day, is nothing but a lie, just as if you said that *AB* was 3000 mm. long after a perfunctory measurement with a draper's yardstick. To affect a higher degree of precision than one can vouch for is just as reprehensible in the one case as in the other. Dates can be determined with reference to a definite calendar by means of authentic documents, or by means of coincidences with other events duly dated, or by a more complicated system of deductions. In every case they are determined within certain limits, and we are bound to state them as correctly as is possible within these limits, the limits themselves being indicated by the choice of units or more explicitly.

I have selected these two examples, lengths and dates, because they are exceedingly simple and yet fundamental. Physical measurements are generally reduced to measurements of length (linear or angular; on the object itself or on our instruments); as to the dates, they are to the historian what spherical coördinates are to the geographer or the astronomer. Nothing in either case could be more fundamental. More technical examples could not provide better illustrations of the argument, for these hit the root of the matter: precision has the same meaning in history as in science, and entails the same obligations.

It is true the errors of historians may remain unnoticed; they cannot be found out as easily as the scientific ones (some of the latter would be almost automatically detected sooner or later); but this does not decrease the historian's responsibility; it increases it.

In physical measurements, the readings of our instruments need correction. If we measure the length of a bar of metal, we must take the temperature into account, and reduce our sundry readings to the same temperature. If we measure the coördinates of stars, we must take into account the aberration of light, the nutation of the earth's axis, the atmospheric refraction, the systematic errors of our instruments, our personal equation. Similar corrections may have to be made with regard to dates; these cannot be compared without having been properly reduced and inserted in the same chronological series. Moreover the date which we read in a book or document may not be correct; it may be vitiated by accidental or systematic errors.

To illustrate. One of the earliest printed editions of Ptolemy's *Geography* bears on its colophon the date 1462 (see Fig. 1). It has been proved that that date is certainly wrong; the real date is probably 1477.[1] With the clumsy Roman numerals such errors were easy enough. If the printed date were true, that edition would be the *princeps*, which it is not. The first dated edition was printed at Vicenza in 1475, without maps; the first edition with maps

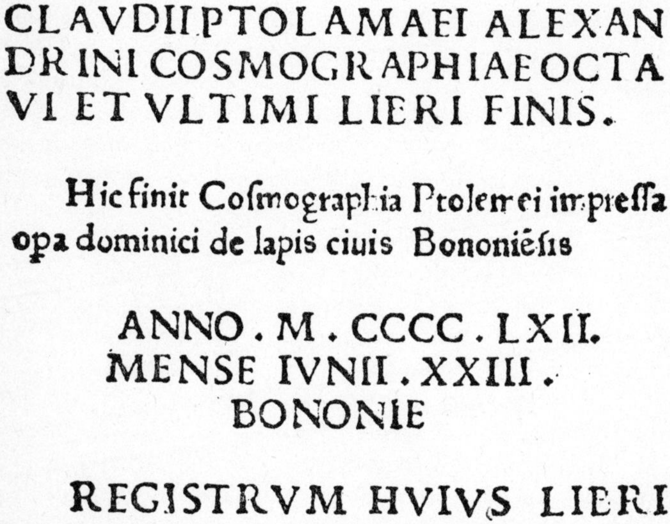

CLAVDII PTOLAMAEI ALEXAN
DRINI COSMOGRAPHIAE OCTA
VI ET VLTIMI LIERI FINIS.

Hic finit Cosmographia Ptolemei impressa
opa dominici de lapis ciuis Bononiẽsis

ANNO . M . CCCC . LXII.
MENSE IVNII . XXIII .
BONONIE

REGISTRVM HVIVS LIERI

FIG. 1. — Colophon of an early edition of Ptolemy's *Geography*, printed in Bologna probably in 1477. The date of the colophon, 1462, is certainly wrong. (Courtesy of the John Carter Brown Library in Providence, R. I., and of Prof. R. C. Archibald).

is the Roman of 1478. In such cases as this, the historian must disbelieve what he sees, and discover more or less painfully the real truth behind the apparent one. He must do with other methods what the physicist is often doing in his own province.

For the period posterior to the invention of printing, the

[1] Sarton, *Introduction*, vol. 1, p. 275, giving 1482 as the date. For the more probable date 1477 see Lino Sighinolfi, "I mappamondi di Taddeo Crivelli e la stampa bolognese della Cosmografia di Tolomeo," in *La bibliofilía*, anno x, pp. 241–269 (1908).

date of the first printed publication of a discovery is generally assumed to be the date of that discovery. However, there are snares for the unwary, and the rule needs many qualifications. For instance, the cautious historian is aware that the dates printed on the covers of periodicals are not always truthful, and when delicate questions of priority are involved these printed dates cannot be accepted as evidence; one must take the trouble of determining the dates of issue.[1]

In some cases the publication of a discovery has occurred *viva voce* in a public lecture or at the meeting of an academy. It may then be necessary to distinguish between that oral publication and the printed one, which may be considerably altered and delayed. These two publications should be considered as two editions of a book, editions which may or may not be essentially different, and of which the first is very limited. A famous example is that of William Harvey, who explained his discovery of the circulation of the blood in a public course of lectures before the College of Physicians in London in 1616, witness his own manuscript lecture notes,[2] and again twelve years later in the *Exercitatio Anatomica de Motu Cordis et Sanguinis*. We must thus distinguish two publications: the first, oral, in 1616 to a limited circle of London practitioners; the second, through a book printed in Frankfort in 1628, to the whole republic of letters. There

[1] Sometimes the date of issue is stated on the cover, but that is not conclusive. The issue may be postponed at the eleventh hour for the sake of an important correction or addition, and the date left as it was. In many great libraries, the dates of reception of individual numbers are often marked by means of a rubber stamp: the date of issue may then be deduced from such dates of reception.

The Société Asiatique of Paris is a great offender in this respect. For instance, the two numbers constituting vol. 235 of the *Journal asiatique* and bearing the dates "juillet–septembre 1934" and "octobre–décembre 1934" were actually received by me in Cambridge, Massachusetts, in March and June, 1935. Scholars might eventually claim on the basis of such arbitrary and misleading dating a priority which they would not deserve.

[2] Charles Singer, *The Discovery of the Circulation of the Blood* (London, 1922; cf. *Isis*, vol. 5, pp. 194–196). See also Sir William Maddock Bayliss, *Principles of General Physiology*, 4th ed. (London, 1924), pp. 666–669.

is no evidence that the first ephemeral publication had any influence; the second being permanent could wait; its influence was felt very slowly and gradually, but irresistibly.

For another illustration of the constant need of vigilance, let us turn to tombstones. We are all cautious enough with regard to the moral judgments included in epitaphs. We know that the 'devoted husband' may have been a libertine, and the 'beloved and respected father' a selfish tyrant, but we are singularly prone to accept the dates carved on stones as correct. Even so great a scholar as the late Percy Stafford Allen, the editor of Erasmus's letters, made the following imprudent remarks apropos of the tomb of Alonso de Fonseca, archbishop of Seville (died 1473), which he visited in the castle of Coca. "Here on his tomb was an authority beyond which we need not seek, the date of his death graven securely on stone."[1] If the tomb was built within the year of death, we may perhaps assume that the date is correctly graven, though the stone cutter might forget one of the many items of a Roman date as easily as the printer; if the tomb was built some time later, as often happened, the causes of error increase. I have already quoted the case of that heroic Transylvanian scholar, Körösi Csoma Sándor.[2] A monument to his memory in Darjeeling, Sikkim, proclaims the following, "On his road to H'Lassa to resume his labours he died at this place on the 11th April 1842, aged 44 years" (see Fig. 2, borrowed from Theodore Duka's biography, 1885, p. 154). Now we know that Csoma was born in 1784; he was then 58 years old at the time of his death, not 44.

The moral of this is that the historian, like his brother scientist, must accept nothing without investigation and verification; he must be forever on the alert, and nurse skepticism in his mind until every suspicion has been dis-

[1] P. S. Allen, *Erasmus, Lectures and Wayfaring Sketches* (Oxford, 1934), p. 188.

[2] Sarton, "Preface to Volume Twelve," in *Isis*, vol. 12, pp. 5–9 (1929). See also *Isis*, vol. 16, p. 180.

pelled. In practice it is obviously impossible to verify everything, but the experienced and wise historian knows when to relax his vigilance and when to intensify it. The difference between careless and careful work is of exactly the same order in the field of history as in that of science. Careless work is easy enough, while every additional precaution increases the difficulty, the slowness, and the tediousness of the endeavor. Even as the good scientist must be prepared to take infinite pains and work over endless trivialities in order to be sure of the accuracy of his experiments, even so the historian must be ready to check and recheck, in as many ways as are available, countless details, each of which may seem unimportant. Unfortunately the law of diminish-

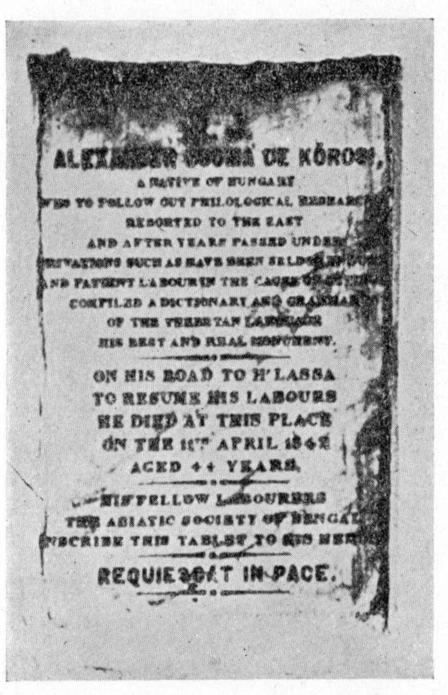

FIG. 2. — Epitaph on the monument to Alexander Csoma in Darjeeling. The implied date of birth is wrong.

ing return applies to historical research as well as to scientific research. Beyond a certain point additional refinement becomes increasingly more expensive, and after a while the expense becomes prohibitive. Long before reaching that limit, the work may become so hard and boresome that its value may be questioned. "Is all that travail worth while?" sighs the fagged scholar in a moment of despair. It *is* worth while, for, in history as in science, inaccurate data are of no value. However cheaply the latter may be bought, their price is too high, while truth, or even the nearest approach to it that can be won, is never too dear, for it is beyond price.

One should, indeed, avoid needless precision, but historians, like scientists, hesitate to draw the line, and for the same reasons. We may not need a high degree of precision for our own investigations, but no one can foretell the degree of precision which may be needed later on for other investigations. Moreover, it is always possible to deduce less precise data from more precise ones, while the contrary is impossible. If we have found the length of a wire to be 1998 mm., we can say later that it was two metres long; but if we put on record a length of two metres and lose the wire, it will remain forever impossible to indicate its length in millimetres. If we know that Copernicus died on May 24, 1543,[1] we can always simplify that statement, and say that he died in May, 1543, or in 1543, or toward the middle of the sixteenth century, which would suffice in many cases, but we could not reverse the procedure. Accordingly it is generally advisable to adapt our precision not so much to our immediate need as to our power of attainment.

The average degree of precision in historical investigations is unfortunately much smaller than that within the reach of the physicist, the chemist, or even the physiologist. The story has often been told of how Sir Walter Ralegh, on receiving I forget how many different accounts of an inci-

[1] For a discussion of this date see Leopold Prowe, *Nicolaus Coppernicus* (Berlin, 1883), vol. 1, part 2, pp. 554–556.

dent which he had witnessed from his own window at the Tower, laughed at the idea of his writing a history of the world. And yet he did write it! No scientist worth his salt has ever abandoned an investigation simply because the attainable precision was too low. Our duty is to be as accurate as we can; it is independent of the absolute degree of accuracy. Our merit is in every case proportional to the ratio of the amount of truth which we discover to the amount discoverable.

In the search after truth, one can never be too cautious or too humble. The good scientist understands that very well in his own domain, but he is likely to throw humility and prudence to the winds when he deals with historical subjects. Now this is intolerable and indefensible. He is not obliged to do historical work; but if he does, he should remain faithful to his own standards of accuracy and honesty, or else be freely castigated and discredited. Careless historical work is as contemptible as careless experimental work, and errors due to the neglect of well known historical methods are as disgraceful as errors due to the neglect of well known experimental methods.

When the scientist realizes all this, and becomes aware of the existence of historical pitfalls similar to the ones wherewith he is familiar in his own laboratory, he is on the road to wisdom, and when he develops a historical conscience as well as a scientific one, his new education may be said to be well begun and his initiation into humanistic studies prepared in the best manner.

* * *

It is obviously out of the question in these few introductory pages to describe our field of study even in outline, but I shall try to give an idea of the immensity of our field and of the complexity and variety of our interests. The historian of science must consider the development of science and technique from the earliest beginnings down to our own days, in all countries, and by people of all races and all faiths; he must consider the development of science at every time and in every place. He must be prepared to extend his investigations as deeply into the past as the emergence of human documents allows, and yet keep his scientific knowledge as up-to-date as possible. To be sure, nobody is supposed to know equally well the development of every science at every time, but the professional historian of science should have some knowledge of the whole field, even as the astronomer, no matter how narrow his special interest may be, is expected to have a general knowledge of astronomy.

The most severe of these requirements is perhaps the one concerning the living and ever expanding science of our time. It is hard enough for the specialist to keep abreast of the steady progress of his own field of study, and yet he is only expected to survey a relatively small sector of the whole horizon. The historian of science should have some acquaintance with the whole field of advancing knowledge, with all the frontiers of science, though he can hardly be expected to enter into technicalities. This obligation rests upon his shoulders because the more he knows of the science of to-day, the better will he be able to appreciate that of the past; early science was once alive, even as our own will soon be dead, and to understand the life of science we must observe it as it grows around us. Could the palaeobotanist understand fossil plants if he were not familiar with living ones? This requirement is difficult enough for the historian of science who has laid a serious scientific foundation to his studies; it is hopelessly beyond the reach of the one who lacks such foundation. Without a sufficient knowledge of modern science, he is unable to understand earlier stages of knowledge, or what is worse, when he believes that he understands, he is almost bound to misunderstand. One of the most pernicious types of error to which a false or shaky knowledge of living science frequently leads is the reading of modern conceptions, such as atomic ideas, energy, evolution, into ancient texts. Nationalist or religious prejudices have often exposed Hindus or Muslims to errors of that very kind, and some western scholars have been caught in the same trap because of their irrational love for the Middle Ages.

* * *

We shall realize in another way the immensity of our task when we try to divide it, as we must, either for study or teaching. Moot points arise at once. How shall we divide the past? How shall we define intervals of equal importance? One might be tempted as a first approximation to give equal importance to periods of equal length, but the result of this would be to give to ancient times an exaggerated importance, and to eclipse the astounding achievements of our own days. Many scientists and philosophers seem to believe that 'real' science dates only from the seventeenth or sixteenth century. Even a superficial knowledge of ancient history is sufficient to prove the falseness and the absurdity of that belief, but it is a demonstrable fact that the progress of science is constantly accelerated,[1] and hence that more and more is accomplished in shorter and shorter periods.

To show how much our perspective of the past would be distorted by an assumption of its homogeneity, let us assume that the total past to be taken into account measures 5400 years — fifty-four centuries, hardly more than two hundred generations, if as many — that is, let us begin our history c. 3415 B.C. Then, on a full circumference, each year would be represented by four minutes, each century by 6° 40′. Our three golden centuries, the seventeenth to the nineteenth, would occupy a sector of only 20°! (Fig. 3). And that assumption, it should be noted, would be very far from

[1] The acceleration *seems* to have diminished about 1882 (see below), but that is another story.

satisfying to our prehistorians. They would say, "3415 B.C.! That is hardly more than yesterday in the story of human evolution! By that time *the main work was already done*, and had been done for ages, for it is the first steps which are the most difficult as well as the most de-

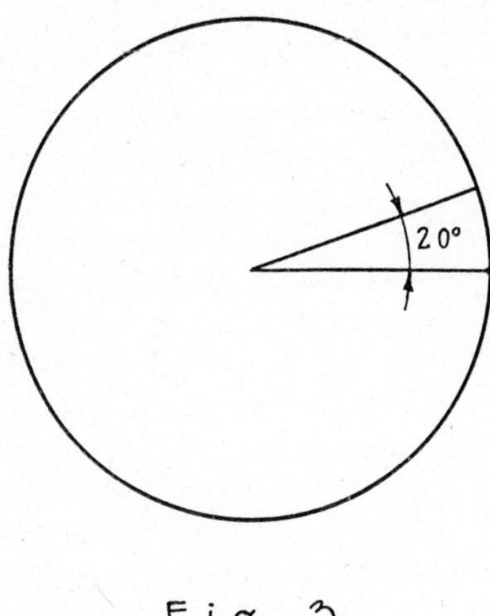

Fig. 3

Sector of 20°, representing three centuries if 360° = 5400 years.

cisive." The history of science must be made to begin with the fundamental inventions: language, drawing, writing, artificial fire, elementary tools, etc. The unknown fire-makers and the inventors of the wheel were the true ancestors of our Edisons and Marconis, and not a bit inferior to them. It required, probably, more genius to invent the first wheel than the latest dynamo. We should begin our survey at least half a million years ago. If we assumed a length of about 648,000 years for the whole record down to now, then on a full circumference each year would be represented by two seconds, and our glorious twentieth century, so full in spite of its youth of tremendous discoveries, by hardly more than a minute, so that it would not appear at all even on a large sized figure.[1]

* * *

I have insisted upon this simply to help in correcting the opposite error of perspective usually made by scientists. Their tendency is to divide the past, not into periods of equal length, but into periods of equal productivity. On that basis the ancient times and the Middle Ages are entirely sacrificed to our more immediate past, when the slow gestation of centuries and millenaries was suddenly followed by an amazing outburst of discoveries.

Let us see how their conception works. A convenient measure is provided by Ludwig Darmstaedter's *Handbuch zur Geschichte der Naturwissenschaften und der Technik* (2d ed., Berlin, 1908), a list of discoveries enumerated year by year from c. 3500 B.C. to 1908 A.D., a total of 5407 years or 54 centuries. I give below the number of pages and the percentage devoted to successive periods:

TABLE I

Period	Pages	%
35 centuries B.C.	28 pages	3%
Cents. I to XV	43	4
XVI	33	3
XVII	55	5
XVIII	117	11
XIX	717	67
Years 1901–08	77	7
	1070 pages	100%

[1] Here is another illustration of the same idea given by J. H. Robinson. "Let us imagine . . . that 500,000 years of developing culture were compressed into 50 years. On this scale mankind would have required 49 years to learn enough to desert here and there his inveterate hunting habits and settle down in villages. Half through the fiftieth year writing was discovered and practised within a very limited area, thus supplying one of the chief means for perpetuating and spreading culture. The achievements of the Greeks would be but three months back, the prevailing of Christianity, two; the printing press would be a fortnight old and man would have been using steam for hardly a week. The peculiar conditions under which we live did not come about until Dec. 31 of the fiftieth year." "Civilization," in *Encyclopaedia Britannica*, 14th ed. (London, 1929), vol. 5, p. 738.

That is, the earliest fifty centuries out of fifty-four (93%) are disposed of in less than 7% of the total space, while the nineteenth century alone (less than 2%) covers 67% of it!

Or, to put it in another way, let us divide the *Handbuch* into sections of a hundred pages each, and see to how many years each of these sections corresponds, in absolute numbers and percentages.

TABLE II

Pp. 1–100, c. 3500 B.C. to 1591 A.D.,	5090 years	94.14%
Pp. 101–200, to 1756	165	3.05
Pp. 201–300, to 1809	53	.96
Pp. 301–400, to 1832	23	.44
Pp. 401–500, to 1847	15	.28
Pp. 501–600, to 1860	13	.24
Pp. 601–700, to 1872	12	.22
Pp. 701–800, to 1882	10	.18
Pp. 801–900, to 1891	9	.17
Pp. 901–1000, to 1901	10	.18
Pp. 1001–1070, to 1908	(7) [1]	.14
	5407 years	100.00 %

[1] The annual rate for this period is about the same as for the thirty years previous.

That is, the first hundred pages deal with 94% of the total time, the first three hundred pages (28% of the book) with 98% of the total time, while the rest of the book (72%) deals only with the last 2% of the fifty-four centuries.

Table II is very interesting, because, if the data upon which it is based are correct, — and they may be considered so at least for a first approximation, — the acceleration in the progress of science did *not* continue to increase after, say, c. 1882. It should be noted in passing that the amazing acceleration in the rate of discoveries and inventions in the nineteenth century was partly mechanical: it was due to the increase in the number of universities, laboratories, and investigators. It is clear that to determine the natural acceleration it would be necessary to make allowance for the artificial acceleration due to the multiplication of centres of research and of experts. However, if the purpose is to measure the acceleration of scientific progress by whatever means, then the figures of Table II do not need such correction.

* * *

It is clear enough that this last picture, which we may call, for short, the scientist's picture, is just as distorted as the former, which we may call the prehistorian's picture. The

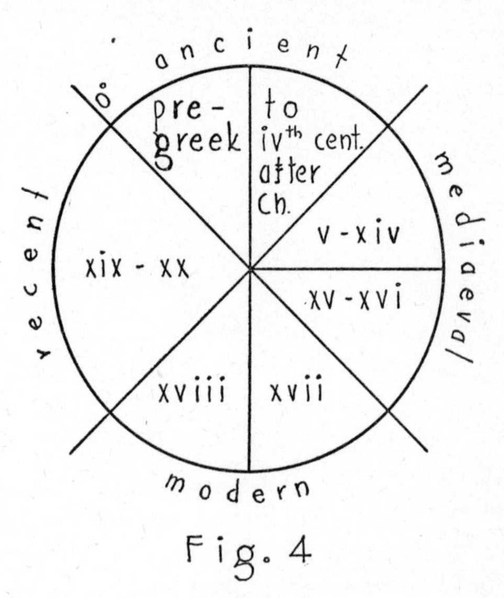

Fig. 4

true picture must be somewhere in between, though a precise determination of it is obviously out of the question. Indeed, there are no accurate means of measuring the value of past achievements. Bearing always in mind that initial efforts are by far the most difficult, and, in general, that equal efforts made at different times should be given very different weights, and that equal efforts made in different countries, or under different circumstances, have also very different weights, our division of the past must be necessarily rough and tentative.

A scholar's intuition may well provide the best answer, the value of his answer being a function of his experience and wisdom. My experience extends over twenty-five years, during which I have been obliged not only to lecture

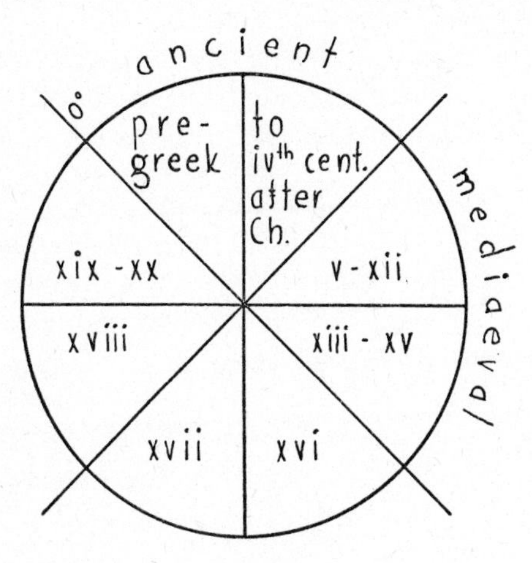

Fig. 5

on a great many subjects, covering among them almost the whole field, but also to examine a very large proportion of the writings devoted to our studies by scholars of almost every civilized country. My own solution may thus deserve to be considered. I have really two solutions between which I often hesitate and which are best represented by the two diagrams above (Figs. 4 and 5). The main difference between these two diagrams lies in the last quadrant, which in Fig. 4 is devoted to recent times only, the last 135 years, while in Fig. 5 it includes the eighteenth century as well. It any case these diagrams should not be taken too literally; they are nothing more than suggestions of how to balance the past against the present.

Many scholars, I am sure, would wish to increase the share allotted to Greek science (only half a quadrant) at the expense of the Middle Ages. They would have good reasons for doing so, for the scientific achievements of the Greeks and of the Hellenistic age were incomparably greater than those of all the centuries following until the seventeenth. It should not be forgotten, however, that in any full account of mediaeval science much space is necessarily devoted to the transmission of ancient science and its discussion. Even as Greek science was preceded by an enormously long period of preparation, which deserves careful study (we give half a quadrant to it in both schemas), even so the Middle Ages was the time of gestation of modern science. It was a new incubation of all ideas inherited from the near and distant past and of newer ideas gradually added to the mixture. The study of mediaeval tradition and invention is not so thrilling as the Greek miracle, but it is very important in another way and very interesting. A relatively large amount of time is needed to explain it, because the story cannot be tolerably complete without a full account of Oriental (chiefly Arabic) science, and the disentanglement of every skein of tradition is very complicated.

The question which we have just discussed, the proper division of the past, is not only of theoretical but also of practical interest, for every professor of the history of science, and every author of a textbook which is supposed to cover the whole field or a large part of it, must decide how he will divide his time or his text. Whenever I have to review a book dealing with the history of science in general, my first test of it is almost always to find out how the author solved this fundamental problem; his solution gives me valuable information with regard to his tendencies, experience, and wisdom.

* * *

Though a teacher of the subject may have to cover the whole field, he cannot be expected to have a first-hand and deep knowledge of every part of it, any more than the professor of general chemistry, let us say, can be expected to have a first-hand knowledge of the whole of chemistry. However, in this case as in others, first-hand and experimental knowledge of one part of the field enables one to appreciate more critically other parts of it, though one's familiarity with them must necessarily remain very imperfect.

One of the best means of obtaining a relative mastery of the whole field is to make judicious cross-sections of it, and every scholar engaged in these studies should try to carry out as far as possible three different types of cross-sections and to remain sufficiently familiar with them throughout life. He should make a special study of:

(1) The development of one branch of science, say, astronomy, or of a smaller field, say, geodesy.

(2) The development of science and learning during a special period, say, the fourteenth century.

(3) The development of science and learning within a certain country, or by the people of a special race or faith, or by the people using a special language. E. g., Italian science, Arabic or Islamic science, the scientific content of post-classical Latin literature.

The well tempered historian of science who has become sufficiently familiar with three such sections might be said to be adequately prepared, though his actual mastery of the subject would naturally depend upon his experience, diligence, and wisdom.

It should be noted that the three perpendicular sections are not artificial. On the contrary, almost every scholar exploring the field with some method would be irresistibly led to the employment of the three kinds. To begin with, his own scientific preparation would introduce the first cross-section. If he had obtained his main scientific experience in mathematics, he would naturally be more interested in the history of mathematics than of other sciences and better equipped to study it. His own nationality or race would determine the second cross-section. If he were an Italian, he would be more interested in Italian science and more competent to deal with it. The reading of old MSS. is always a good deal easier for one who has a native knowledge of the language than for any other scholar, however well the latter may fancy he knows it. The use of libraries, archives, and collections, and the tracking down of documents is also much easier and more fruitful for a native or citizen of the country involved than for a foreigner. The choice of the horizontal cross-section is perhaps a little more arbitrary, though it would be partly determined by the two other sections. A student of the history of mathematics would be easily fascinated by the golden age of Greek mathematics, or by the seventeenth century, and would desire to know other aspects of those privileged periods than the purely mathematical; or if he were a Muslim, an Arab, or an Arabist he would wish to explore more completely the age when Arabic culture was supreme. These three cross-sections, it should be noted, are never independent; each includes projections of the two others and completes them (Fig. 6).

* * *

What I have said thus far illustrates not merely the immensity of our field, but also its endless and bewildering variety. Such subjects as, let us say, Greek geometry, Chinese alchemy, Hindu astronomy, seismology, and internal secretions are as different as can be; they imply not only different equipments but different attitudes of mind.

So large and complex is the field, that though I have explored it during a quarter of a century in almost every conceivable direction, I would not dare say that I know it. The difficulties are caused partly by its complexity and infinite

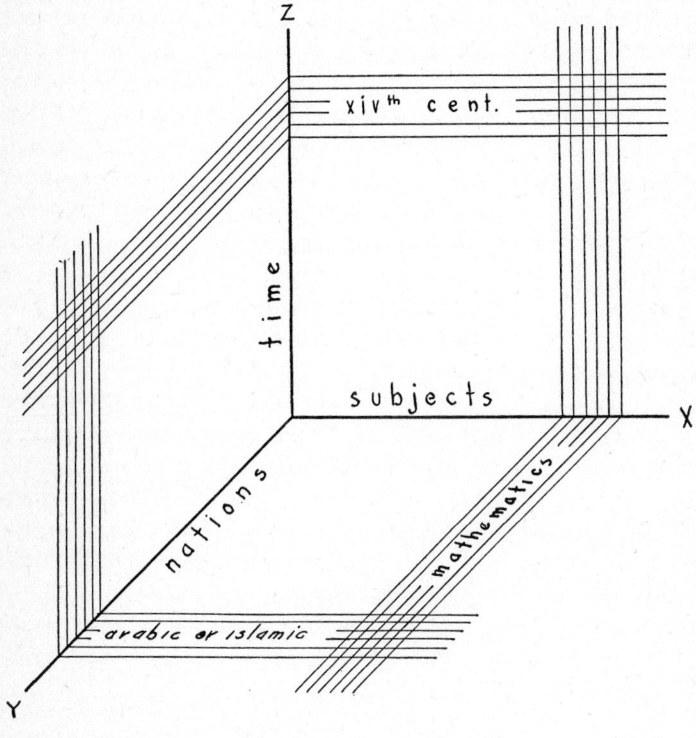

Fig. 6

variety, partly by the lack of tools, due to the very youthfulness of our discipline.

It is worth while to stop here for a moment and consider our lack of tools. Though the great men of science are in their own unobtrusive way the leaders of mankind, and must be counted among its foremost benefactors, there are relatively few full-sized biographies devoted to them, and in many cases we even lack short ones that are, at the same time, accurate. There are many collections of such biographies, but the bulk of them are meretricious little books, which may satisfy the average reader, but are of no value for the historian. Even at their best, as in Cuvier's *Éloges historiques*, they do not measure up to the standards of good biography obtaining in other fields.

Take another type of monograph, the study of separate questions, as, e. g., the discovery of specific and latent heats.[1] To be sure, most of these studies are contributions to the history of particular branches of science, but the main questions of each science concern the whole of science. Problems of specific and latent heat are not the affair of physicists only, but also of chemists, geologists, astronomers, and physiologists. The best way to realize our lack of such fundamental monographs is to take any general textbook. The experienced reader will soon notice that there is hardly a page or paragraph which does not require, for the sake of completeness and accuracy, the preparation of a monograph which does not yet exist. Strictly speaking, those textbooks should not be written until the necessary monographs were ready as a foundation; strictly speaking, the elementary books should be the last to be composed; but their premature publication cannot be helped, for we need them right away; at every level of our growth we need syntheses to guide our new analytic efforts.

The point to keep in mind is that the work remaining to be done is enormous. This is not surprising when one realizes, on the one hand, the immensity of the task, and on the other the fewness of the workers, as compared with the legions of other historians engaged in the study of political or economic history, the history of religions, of art, of lan-

[1] This particular topic comes to my mind because it has recently been treated with great care by Douglas McKie and Niels H. de V. Heathcote (London, 1935; cf. *Isis*, vol. 25, p. 227).

guages and literatures, of arts and crafts, in short, of everything except of that which is the backbone of culture: science.

As one question after another is investigated down to bed rock, as better textbooks and treatises are gradually produced, and syntheses of larger and larger fields made at higher and higher levels, it will become easier to know the whole subject. The total knowledge will be very much larger and more precise, but it will be so clarified and so well ordered that its comprehension will be simpler.

* * *

We shall understand this better if we make a distinction between what I should call real knowledge and potential knowledge. For example, if I owned a book containing a list of all countries and cities of the world with their positions, areas, and populations, all according to the latest and most authoritative sources, if there were such a gazetteer at my elbow, I could say that I had a potential knowledge of its contents, for pertinent information of the best kind could be extracted from it immediately, as one would extract it from one's memory. In fact, such a book might be considered an extension of one's memory, though far more reliable than the best memory could ever be. This is an extreme case; passing to the other extreme, it is equally true that the possession of the best textbook on the theory of functions would not give me *ipso facto* any potential knowledge of the subject. If I had made years ago a thorough study of it, and since then forgotten most of it, my real knowledge would be dwindling to zero, but the textbook might replace it by a potential knowledge. In this case, however, the potential knowledge would not have been obtained gratis, as in the first one, but only at the price of long antecedent efforts.

The branches of science and learning might all be arranged in a single row according to the degree of availability of potential knowledge, the ones where that degree is lowest, the mathematical branches, being placed at the extreme left, the ones where that degree is highest at the extreme right. Happily for us, the historical sciences would find their proper place toward the right end. In other words, a fair amount of historical knowledge is immediately available to any intelligent man owning the proper tools. The special tools of the historian of science are not yet ready — or, at least, very few of them —, but they can be made, and will probably be available to the next, or next but one, generation of scholars. When his own good tools are finally ready for him, and he has been trained in their use, a large amount of potential knowledge will be open to him. To be sure, a treatise on the history of mathematics will not be immediately helpful to the non-mathematician, but it will extend with little effort the potential knowledge of the historically-minded mathematician.

* * *

To return to our subject. However difficult the determination of scientific facts of the past may be, e. g., the definition of a discovery and its correct placing and dating, their interpretation and final appreciation are exceedingly more difficult, and especially so because in this case historical rules and methods are of little avail; what is needed is the total scientific and historical experience of the judge, the wisdom and subtlety of his mind, the charity of his heart. Here is a fact. On March 13, 1781, William Herschel discovered a "curious either nebulous star or perhaps a comet" which turned out to be a new planet, Uranus. What does that mean? In order to appreciate it, one must bear in mind the history of astronomy down to that time and realize that this was the first planet to be discovered within the memory of mankind! One must bear in mind the circumstances of Herschel's life and activity. How did he discover Uranus, and why did nobody else discover it before him? What were the implications, for him and others, for us, of that discovery? The statement of the discovery is almost meaningless, unless all these explanations are implicitly available to the reader, or explicitly stated to him. A bare fact is nothing, though its potentialities may be endless; meaningless to one person, it opens up to another a new chapter of science or life, and provides the kernel of a tragedy. The historico-scientific facts are the stones of the building which the historian of science is building; if accurate, they are invaluable — history without them is nothing but a vision, a dream, a bubble, a shadow, and even less, the shadow of a shadow; yet they are worthless unless placed in their proper sequence and position, each one being explained and justified by all the others.

* * *

The search for new historical materials is itself a task of great complexity, to the extent that only a part of it, and that not the most important, can be taught. However, the teachable part should be learned; the delicate methods involved are explained in historical seminars or in special textbooks. We cannot insist upon that. When the historian of science has finally succeeded in tracking down some new document, by the use of such methods or his own shrewdness, or both, his next task is to criticize it, externally and internally, and to make the best use of it. If he be dealing with an early text, which may be represented by more than one MS., he must first 'establish' it, that is, he must determine the most probable text. Of course, the establishment of a scientific text, its external and internal criticisms, are not different from those relative to any other texts. The methods are exactly the same, but the internal criticism, i. e., the interpretation of the contents, will necessarily require an adequate knowledge of the scientific ideas involved and of their implications. To appreciate a scientific text fairly, it does not suffice to know how much of it is true from our more advanced point of view, we must find out how much of it was considered true at the time of its composition, how much was true or plausible within the framework of contemporary science, how much was new. What were its sources? What was its influence? Two questions which seem very simple, yet either of which may involve endless investigation. We must reconstruct link by link the sequences of ideas leading to that text, and the new sequences leading from it. We must place that text as exactly as possible within the complicated network of our traditions.

This is, again, a set of problems with which historians are familiar enough, and scientists unfamiliar. Modern scientists hardly think of tradition, because it is now insured by various competitive agencies. The average scientist need not worry about it; as far as he is concerned, the tradition is almost automatic. It might be interrupted here or there by some physical or social disaster; it could hardly be stopped everywhere and forever. A geologist who takes the trouble of examining a few geological journals of various countries knows almost everything that is being done within his field and is worth knowing. Should an exceptional discovery or a startling theory be published — say, the Wegener hypothesis — he would hear of it not only once, but a hundred times; his own newspaper would speak of it; indeed, he would have to stop his ears not to hear of it.

Conditions were very different before the existence of scientific journals; they were worse still before the invention of printing. The tradition of knowledge in ancient times and even in the Middle Ages was exceedingly casual and problematic; it might be jeopardized at any time by a social catastrophe if not by a trivial accident; at best it was very slow and irregular. Indeed, when one thinks of all the vicissitudes of ancient history it seems almost miraculous that so many early texts — say, abstruse mathematical treatises which could not interest many people and were never represented by many MSS. — have come down to us. And those that have survived have reached us with almost unbelievable indirectness. For example, the treatises of Hippocrates, Aristotle, Galen, originally written in Greek, were translated into Syriac, then into Arabic (or directly into Arabic), then into Hebrew and into Latin (or into both), then into our own vernaculars. Each of these stages entailed errors of omission or commission, each translation was somewhat of a gamble, each edition was represented by divergent manuscripts, and so forth.

When one visualizes all these hazards and perils, is it not astounding that we are able to read and enjoy to-day a treatise which Archimedes wrote twenty-two centuries ago?

It is true that the ideas which the ancient treatises contained may have trickled down to us in countless other ways. The historian's task is then not simply to determine the contents of Ptolemy's knowledge, but to trace the tangled channels through which that knowledge has reached modern astronomers such as Copernicus, and how its mixture with other intellectual streams of equally complicated ancestry has helped to constitute, very gradually, our own conceptions. Ptolemy's works are one thing, their tradition quite another, both indispensable. When we are told that our energy is derived from the sun, we are not satisfied until we are shown how that energy is communicated. It is a very complicated story; the history of the Ptolemaic tradition is equally complicated and fascinating.

* * *

As the historian is expected to determine not only the relative truth of scientific ideas at different chronological stages, but also their relative novelty, he is irresistibly led to the fixation of *first* events. "So-and-so was the first to do this-or-that." "This was the first treatise dealing with . . ." This never fails to involve him in new difficulties, because creations absolutely *de novo* are very rare, if they occur at all; most novelties are only novel combinations of old elements, and the degree of novelty is thus a matter of interpretation, which may vary considerably according to the historian's experience, standpoint, or prejudices. At any rate the determination of an event as the 'first' is not a special affirmation relative to that event, but a general negative proposition relative to an undetermined number of unknown events. It is always risky, yet when every reasonable precaution has been taken one must be willing to run the risk and make the challenge, for this is the only means of being corrected, if correction be needed.

* * *

We remarked a moment ago that scientific tradition has now become almost automatic; we may be sure that scientific novelties will almost inevitably be known within a short interval of time to the scientists working along the same lines. However, the problem of tradition reappears in another form, which may leave the pure scientist curiously indifferent, but touches the humanist on the raw. How is scientific knowledge transmitted, not to the specialists (that is easy enough), but to other scientists, and, most difficult of all, to non-scientists? How is scientific knowledge to be taught to children, how is it to be diffused to the educated world? And, most pregnant question of all, how can scientific methods and points of view be inculcated in people's minds?

Most educators, not to speak of non-educators, have false notions on the diffusion of scientific knowledge. It is very important that every man should be made to realize the immensity of the universe, and the inverse immensity of atoms, but does it really matter how many know the latest size of the universe, or are able to describe accurately the latest atomic models? Most of the newest knowledge, the kind that is featured in newspapers, is of little concern to all but specialists. On the other hand, it is of fundamental importance that every educated man and woman should be made to understand the methods of scientific research, and be trained to love truth irrespective of his interests and prejudices. There is nothing impossible in that; that is, it should not be more difficult to teach people not to lie, than not to steal. The principles and methods of science could be taught any sound-minded person with only a fraction of the effort that is now misused to teach him the newest particularities which he does not really need. The teaching should be concentrated on the facts and theories which have survived the controversial stage and have (or should have) become part and parcel of the common knowledge of educated men all over the world. The experimental proof and the explanation of those facts and theories (either in the historical order or otherwise) would serve to illustrate the methods and spirit of science. The republic requires relatively few scientists, but it will not be healthy until a majority, or at least a large minority, of its citizens have been trained to consider every-day problems, and to judge political and social controversies, with a certain degree of objectivity and disinterestedness.

* * *

The mention of scientific methods suggests another aspect of our investigations: the study of the development of science from the logical point of view. The fact that the logic of science is largely casual and retrospective does not matter. It remains of great interest to discover inductively the logical sequences or the logical solutions of continuity in the arguments and activities which have led mankind from one discovery into another, and from each scientific level up to a higher one indefinitely. We can thus reconstruct, or help to reconstruct, as it were, the development of the human genius; that is, not the intelligence of any single man or group of men, but that of mankind as a whole. It is clear that this field of inquiries provides a necessary complement to purely historical investigations, and that it is in itself so large that a scholar's life would be too short to explore it completely.

However averse the historian may be to philosophical and logical discussions, he cannot eschew them altogether, because he cannot properly appreciate the value of discoveries without them. To illustrate this let us consider for a moment the idea of *experimentum crucis*. One calls a crucial experiment one which enables him to choose between hypotheses which are mutually exclusive by proving that one of them is right and that the other must be wrong. The classical example is that concerning the wave theory of light and the emission theory. There was much discussion in the first half of the nineteenth century as to which of those theories was right to the exclusion of the other. The latter had been favored by Newton and was generally accepted by the triumphant Newtonians; the first had been brilliantly but incompletely explained by Huygens, and after a century of neglect it had been revived by Young and almost completely vindicated by Fresnel. A few years after the latter's death, Sir William Rowan Hamilton in the course of his analytical development of the wave theory was able to predict the existence of a very rare kind of refraction, the conical refraction, which had never been observed. Hamilton's mathematical prediction of 1832 was verified experimentally by Humphrey Lloyd in the following year.[1] The wave theory seemed to be established on an inexpugnable basis, and yet some of the defenders of the emission theory refused

[1] For more details see G. Sarton, "Discovery of Conical Refraction" (*Isis*, vol. 17, pp. 154–170, 1932), including a facsimile of Lloyd's paper.

to capitulate. It was then that Arago invented an ingenious *experimentum crucis*. If the emission theory is correct, the speed of light must increase with the density of the medium; if the wave theory is correct, the speed must decrease with the density. Hence if we could measure the speed of light in air and water, the comparison of the results would tell us which theory was true. The extremely delicate experiments which he suggested were not realized until fifteen years later by Foucault, who proved that the speed of light is smaller in water than in air, and thus "that the emission theory is incompatible with the reality of the facts" (these are his own words).[1] This seemed to be decisive and final. There is no doubt that the emission theory was incompatible with the facts which he dealt with, but it did not follow that the wave theory was compatible with every other fact. To make a long story short, the study of black-body radiation revealed facts which were incompatible with the wave theory, and led to the formulation of the quantum theory by Max Planck in 1900.[2] Arago's *experimentum crucis* had simply proved that the wave theory was more complete than the emission theory, but it had not proved and could not prove that it was absolutely true.

It is clear that no experiment can be really 'crucial' until

[1] It is well to recall the three steps leading to that conclusion.

Charles Wheatstone, "An Account of some Experiments to Measure the Velocity of Electricity and the Duration of Electric Light" (*Philosophical Transactions*, 1834, pp. 583–591, 2 pl.). Explaining the method of rotating mirrors.

Arago, "Système d'expériences à l'aide duquel la théorie de l'émission et celle des ondes seront soumises à des épreuves décisives (*Annales de chimie et de physique*, vol. 71, pp. 49–65, 1839). Communicated in 1838. Arago explained the principle of the experiments and suggested their realization by means of Wheatstone's method.

Léon Foucault, "Sur les vitesses relatives de la lumière dans l'air et dans l'eau" (*Annales de chimie et de physique*, vol. 41, pp. 129–164, 1854). Realizing Arago's idea.

[2] For a good summary of this question, see F. K. Richtmyer, *Introduction to Modern Physics*, 2d ed. (New York, 1934), chapter 7 (*Isis*, vol. 24, pp. 172–174).

we are sure that our analysis of the possibilities is exhaustive, and this implies an omniscience which is hardly within our reach. Does this mean that the so-called crucial experiments are futile? Far from it: they help us to clear the ground; they enable us to continue further (if not to complete) the logical analysis of a set of problems, and incidentally to study the development of competitive theories and the reactions to them of different minds. The story of the Arago-Foucault experiments will always be one of the most beautiful in the history of mankind.

* * *

Another important concern of the logically-minded historian is the study of the interrelations between the different branches of science and of the gradual creation of new links between them. How did the progress of one science affect the progress of others? The methods which had been developed in a particular field, how were they applied, *mutatis mutandis*, to another, how could they conceivably be applied to others still? I am not prone to exaggerate, as Ostwald did,[1] the heuristic value of the history of science, that is, its value in helping scientists to make new discoveries, but whatever such value it may and does possess, it is clearly the trained logician rather than the pure historian who will bring it out for us, and this may well become the most valuable part of his work. He it is who should analyze for us the logic and psychology of inventions, their concatenations, their influences of all kinds upon one another. This has not yet been done with sufficient thoroughness and on a sufficiently broad basis so that we may judge whether this approach is as fruitful as it may seem, but the effort is worth

[1] Wilhelm Ostwald went so far as to say that the history of science is nothing but a method of research for the increase of scientific conquests. See his paper "La science et l'histoire des sciences" in *Revue du mois*, vol. 9, pp. 513–525 (1910).

making. The results are bound to be very interesting, even if they do not turn out to be as immediately profitable as one might wish. It will never be possible to replace by systematic deductions, or by mechanical applications of old tricks to new problems, the happy intuitions out of which discoveries are born, but the intuitions may be guided and stimulated by such means, and genius itself may be enabled to soar from a higher level and with greater assurance.

* * *

I have already alluded to the psychology of scientific discovery, for, in practice, it is hardly possible to separate psychology from individual logic. However, in contrast to the logically-minded historian, there is one who might be called psychologically-minded, who is interested not only in the genesis of discovery in the individual mind, but in the whole intellectual and emotional make-up of the scientist. How does this scientist compare with another, as a man, or with other men? How was his temper affected by work, rest or play, by success or failure? How was he influenced by his social environment, and how did he influence it? How did he express and reveal himself, or fail to reveal himself? What was the quality of his spirit? His love of truth, his love of beauty, his love of justice, his religion, to what extent were they developed? Or was he indifferent to the world around him, blind to everything except the narrow field of his research? Not only the psychologist but the humanist tries to answer such questions, and others innumerable.

Is this not natural enough? We have some degree of interest in every man and woman whom we approach near enough. Should we not be even more interested in those men who accomplished more fully the destiny of the race? I read this morning in the paper that a man called John O'Brien died suddenly in Boston while he was watching a wrestling match. I have no interest whatever in wrestling, and yet that event shocked me and aroused my curiosity. What did he die of? His was probably a heart case, and the wrestling excited him overmuch. I have no trouble in understanding that, and my sympathy goes out to him, for I have been deeply moved time after time while I was contemplating my fellow men wrestling not with other men but with nature herself, trying to solve her mysteries, to decode her messages. I have been excited by their tragic failures as much as by their occasional victories and triumphs. The same instinct which causes sport-lovers to be insatiably curious about their heroes causes the scientific humanist to ask one question after another about the great men to whom he owes his heritage of knowledge and culture. In order to satisfy that sound instinct it will be necessary to prepare detailed and reliable biographies of the men who distinguished themselves in the search for truth.

Such biographies are interesting not only in themselves, but as materials for the study of man. There is as much variety among scientists as among other people high and low. Their motives may range along the whole gamut from utter selfishness to utter selflessness, and along the whole gamut of every other passion. Their manners and customs, their temperamental reactions, differ exceedingly and introduce infinite caprice and fantasy into the development of science. The logician may frown but the humanist chuckles.

Happily such differences are more favorable to the progress of science than unfavorable. Even as all kinds of men are needed to build up a pleasant or an unpleasant community, even so we need all kinds of scientists to develop science in every possible direction. Some are very sharp and narrow-minded, others broad-minded and superficial. Many scientists, like Hannibal, know how to conquer, but not how to use their victories. Others are colonizers rather than explorers. Others are pedagogues. Others want to measure everything more accurately than it was measured before. This may lead them to the making of fundamental discoveries, or they may fail, and be looked upon as insufferable pedants. This list might be lengthened endlessly.

From the humanistic point of view, every detail in a scientist's life is or may be interesting, because that life is one of the parts in a great tragedy — we might call it the basic tragedy of mankind —, the struggle for knowledge. From that standpoint, it will never suffice to state a man's discovery; one must explain how and why he made it, and why it was he who made it, what idiosyncrasies guided or handicapped him, and so forth. Every human aspect must be considered, because this is not simply a scientific matter but a human one. One must examine his whole behavior, his ways of searching, of finding, of checking and rechecking, and finally — most illuminating of all information — his ways of expressing himself, in short, his style. "Le style, c'est l'homme."

Moreover, one cannot judge him fairly unless one is prepared to consider his achievement not only from the point of view of modern science, but also from the point of view of the knowledge obtaining in his time, and of his own education and experience. He must be seen in his own environment, and also outside of it. We must try to weigh how much he may have been helped or hindered by all kinds of social influences, most of them irrational or at least non-scientific.

To understand fully the human side of science, we must think of scientific achievements as we do of artistic achievements. Indeed they are, at best, of the same kind, though the techniques are of course very different, and even more so the intellectual attitudes. To be sure, in the scientific field even more than in the artistic one, there is a great deal of work which is so commonplace and repetitious that even a saint could hardly lift it up to an inspired level; there cannot but be many 'hewers of wood and drawers of water,' but between these and the great leaders there is an infinity of steps. Discoveries are to be judged not only in abstract terms but in human ones. The first great extension of our universe by Herschel was more startling and more moving than the periodic extensions about which we read in the daily papers ever and anon and which we have gradually come to expect.

The fundamental distinction between scientific efforts, for the humanist, is that some are heroic while others are not. That distinction cuts across all the others; it is independent of scientific value, of method, even of morality. Some men of science have cheered the whole world with their heroism, while others have done greater things, but in a smaller way, without grace and without beauty. We cannot exaggerate the significance of heroic efforts in the field of science, just as in every other field wherein men compete, for these efforts are the salt of life. If the historian keeps his eyes and his heart open for heroism, it will be easier for him to discount the false values created by success, I mean, the kind of success which is bestowed by a public opinion as fickle as it is uncritical. The record of such success (or failure) may have some social interest, but otherwise it is almost irrelevant, for what matters above all is what a man did and was, not what his contemporaries thought of him. Not all their praise, and all the honors heaped upon him, could add an inch to his stature; nor could their disapproval, neglect, and contempt subtract an inch from it. It is the historian's sacred duty to correct in the light of experience the vagaries of contemporary opinion, and to try to replace blind prejudice with well informed equity. It is his duty to reveal the motives and circumstances which cause one man to be great in spite of his failure, and another to be small in spite of his success. It is perhaps less important to point out the vanity, meddlesomeness, or cupidity of one man, than the generosity, humility, and equanimity of another; in other words, it is more useful to insist upon the good qualities of men, when we find them, than upon their shortcomings, but above all we must celebrate heroism whenever we come across it. The heroic scientist adds to the grandeur and beauty of every man's existence; the complacent, humdrum, commonplace scientist, and the meretricious, or simply the tepid, however startling his discoveries may happen to be, does not inspire us humanists and artists, but leaves us very much where we were. We may perhaps overlook him in our account, though not his discoveries, while the hero's life is in itself an artistic creation, an inexhaustible source of joy and happiness.

* * *

The history of science was and should be written from the many standpoints which I have indicated, and perhaps from still others. It is an immense subject which cannot be contemplated and illuminated from too many angles; each particular survey will add something to our knowledge and to our pleasure. Whichever the historian's point of view, he will not deserve the esteem of his colleagues if he does not do his task honestly, and, within the limits of his purpose, as completely as he can.

Such a remark might seem superfluous. It would certainly be so if we were dealing with other subjects. In fact, it would be almost impertinent to say to a zoölogist that he must do his work as honestly and thoroughly as possible, in order that others should not be obliged to do it again! Yet it is very necessary to say that very thing to the scientist who has historical investigations in view, for he seldom realizes that his own scientific methods and standards apply as rigorously to such investigations as to his own. On the contrary, he is likely to consider his historical interlude as an escapade, of which he is at one and the same time proud, because, being out of his element and having thrown his own standards to the winds, he has no others, and ashamed, because he is consciously or unconsciously aware of doing something wrong. This last feeling may lead him to make fun of his own efforts and play the clown, all of which is futile and painful to behold.

Let us repeat once more that the genuine historian (as opposed to the journalist, the hack, or the clown) should proceed exactly as every other scientist. He must determine as accurately as possible the knowledge already available on the topics he is intending to investigate; carry on his investigations with precision and thoroughness, making full use of the requisite methods and taking every precaution to avoid the various pitfalls; and, finally, publish his results in a straightforward manner. This does not mean that his account should be ugly or dull. It should be as well written as possible, but without irrelevancies or incongruous ornaments.

Whatever their task and purpose, the honest scientist and the honest historian try to accomplish it with some finality. They are ready to take endless pains in order that their efforts need not be repeated, but that the results may, on the contrary, be used by their successors as a starting point for new efforts. Thus is the progress of science made possible.

Most of the historical work done by scientists untrained as historians is published without means of verification, that is, with insufficient or imperfect references, and with so little accuracy that it is useless for later scholars. This or that statement may be right or not, one cannot tell; everything is spoiled by promiscuity of good and bad, and sometimes the whole is distorted by the author's fancy or whimsicality, or perhaps his very striving for artistic effect. Such work is obviously a waste of time for all concerned. It can add nothing to our knowledge. It rather debases the knowledge already available by diluting it with a mass of irrelevant and doubtful information whence it cannot readily be retrieved. It is truly a degradation of intellectual energy.

The ambition of the historian, like that of every other

scientist, is to increase or improve available knowledge. There is no reason for hurry, but pains should be taken to prevent slipping back, or else our work becomes hopeless. We must exert ourselves to the utmost in order that our successors may accomplish their task with greater ease and precision, and come closer to the truth than we can. Poincaré elaborated his *Méthodes nouvelles de mécanique céleste* [1] because the approximation sufficient in his day for the computations of astronomers would be no longer sufficient a few centuries later, and he once declared,[2] "We are sometimes happier to think that we have saved a day's work to our grandchildren than an hour to our contemporaries." That is the true spirit: no pains are grudged if they bring us nearer to the goal, however distant it may be. We are not in a hurry, but we must go steadily forward, not backward.

One might apply to historical work, and more particularly to our own researches, where such advice is more needed than anywhere else, the wise remarks of Albert Bayet: [3]

> On conçoit maintenant la signification morale de toutes ces règles minutieuses qui président au travail scientifique. Si le savant s'impose à lui-même tant de précautions rigoureuses, c'est qu'il n'entend pas seulement atteindre la vérité: il veut la rendre manifeste à tous. Il ne lui suffit pas d'être cru: il entend ne l'être que sur preuves sonnantes et trébuchantes. Ces démonstrations exactes, ces vérifications laborieuses dont s'impatiente parfois l'esprit de finesse ou l'esprit poétique, sont la forme la plus haute de l'altruisme: elles impliquent et le désir de s'accorder avec autrui sur les choses essentielles et le désir que cet accord ne soit pas un accord de surprise, un rapprochement passager, mais bien l'expression solide d'une communauté réelle. Le respect et l'amour de l'homme pour l'homme sont l'âme de la recherche scientifique, parce qu'on ne peut pas faire aux autres un don plus précieux que de leur offrir une vérité qu'ils feront leur et qui les fera s'unir par ce qu'il y a en eux de plus haut.

---

[1] See vol. 1 (1892), introduction.

[2] *Science et méthode*, p. 34.

[3] *La morale de la science* (Paris, 1931, pp. 65–66; cf. *Isis*, vol. 19, pp. 241–245).

Alas! It is not enough for the historian to eradicate errors, he must prevent their recurrence. It happens but too often that they are reintroduced by men of science, who use the prestige honestly obtained in another domain for the unconscious dissemination of errors in our own. What would you think of the husbandman who would spare no trouble to root out weeds in his own fields, and would then sow them in the fields of his neighbors? That is exactly what has been done, and continues to be done, by very distinguished scientists. If the movement which I am leading had no other result but to prevent that especially ugly kind of carelessness and selfishness, and to put an end to the degenerescence of historical truth caused by it, I should already feel well repaid for my labor.

* * *

Many techniques are involved in our investigations, not only the ordinary scientific and historical techniques, but others less usual and more absorbing, if not more difficult. For example, the historian of mediaeval science, that is, he who wishes to make a comparative survey of science and learning in the Middle Ages, must be prepared to study Arabic, as a great many scientific books were written in that language. That knowledge of an Oriental language which would be somewhat of a luxury for other mediaevalists is almost a necessity for him, a necessity which entails a vast amount of work.

The study of an additional and difficult technique, such as that of palaeography or an Oriental language, is an excellent training for any scholar. It gives him the exhilarating feeling of coming closer to rare sources and being able to drink from them; it creates in him a sense of independence and mastery; it ought to lift him high above the temptations of clap-trap and irresponsible historical writing. Just as we should have less confidence in a professor of general physics who had not mastered any of the special techniques of physical research, just so we are less inclined to trust a historian of science who is too much of a theorist and who has not experienced time after time the tribulations of defeat and the joys of victory in the digging out of facts from the bed rock.

Every historian must be trained to overcome certain classes of technical difficulties, and, in a way, the harder these are, the better. Nevertheless it is well to remember that this training introduces new perils. It is possible to teach the most difficult techniques and the most esoteric methods to people of mediocre intelligence, but it is impossible to impart to them the wisdom which is needed to make the best of any technique and to master it truly, instead of being mastered by it. In too many cases the technique ceases to be a means to an end, but becomes an end in itself, and that is very silly.

At the end of their excellent textbook on statistics, Professor Harold T. Davis and W. F. C. Nelson give the following warning, which fits my argument as well as their own.[1] When the student has mastered a definite technique, what then?

> The danger [say Davis and Nelson] is that he will over-estimate rather than under-estimate the value of this equipment. Statistical methodology is no magical, or even mechanical, instrument that automatically grinds out valid conclusions and allows the suspension or avoidance of personal judgment. Indeed, it may be said flatly that a statistical conclusion is no better than the judgment of the statistician who produced it. Knowing what tool to employ is just as important as knowing how to employ it. The second can be taught, but the first must be learned. The novice will tend to think that the more high-powered his methods the more cogent his analysis. This is not at all necessarily true. A scatter diagram may well yield more information than a correlation coefficient. The fact that the latter may be carried to several decimal

---

[1] *Elements of Statistics* (Bloomington, Indiana, 1935), p. 334; cf. *Isis*, vol. 25, p. 279. My reference to a book on statistics is not so arbitrary as it may seem. See my "Quetelet," in *Isis*, vol. 23, pp. 6–24 (1935).

places gives a spurious appearance of accuracy, while it may really be concealing such facts as that the relationship is curvilinear or that some of the observations are evidently grossly distorted. In such a case, the apparently crude method is really enlightening, the apparently precise method is really deceptive. Very often a free hand curve drawn through a graph will tell as much about the trend as will ever be revealed by logistics or quintics. Again, the methods may be too refined for the data.

Much intellectual mediocrity can be and actually is concealed by some technique sufficiently recondite to discourage outside criticism, even as social conventions can easily mask the lack of individuality, or religious rites provide the best of screens for moral unfitness, and even for iniquity.

This is as true for science as for history. It is pathetic to think of the efforts made in hundreds of laboratories to realize with the most complicated and forbidding apparatus the most futile experiments.

Parturiunt montes, nascetur ridiculus mus.

The danger of pedantry and of unbalanced technicality will always be with us, because mediocrity is far more common than intelligence, not to speak of genius, and because the number of fools is always large. Moreover, it is well-nigh impossible to say where pedantry begins. We hardly dare interfere with the excessively complicated experiments of an unwise scientist, because it is impossible to appreciate the situation with fairness without identifying ourselves, as it were, with him. We may suspect that his experiments will be futile, but we are not sufficiently sure of it to stop him. Pedantry cannot be inferred from the subject. A man may spend his life compiling a dictionary or a mathematical table and yet not be a pedant; or he may seem to be very catholic in his tastes and yet be a pedant at heart; some of the worst of the tribe are found among self-styled poets and artists.

The fundamental difference between creative scholarship and pedantry lies in the power of selection which wise men have and pedants lack. Now this brings science and art very close together, for right selection is the essence of art as well as of science.

The artist cannot reproduce every aspect of nature or realize every dream of his mind; he must choose, choose, choose. Even so the scientist cannot study every fact and attack every problem; he must choose and choose and choose again. His activities are continually dominated by the need of selection; they may be suddenly exalted by a wise choice, or jeopardized, even nullified, by a wrong one. Genius in science as well as in art includes, as one of its essential elements, that uncanny quality, the ability to select the most characteristic lines or colors, melodies, or harmonies, or the salient fact, the fertile problem, the 'crucial' or enlightening experiment. Granted that selection is even more fundamentally and continually important for the artist than for the scientist,[1] that is, that artistic creation is far more arbitrary than scientific creation, the difference between them is quantitative rather than, as is generally believed, qualitative.

We must try to impart difficult techniques and rigorous methods without making pedants. It is true the problem is largely solved for us before we try to solve it ourselves, for some men are born pedants, and whatever technical skill is given to them will only feed their pedantry; just as some people are born hypocrites, and religious training can only make them worse. However, it is well to know the dangers

[1] Einstein once remarked that the infinitesimal calculus would certainly have been discovered even if there had been no Newton and no Leibniz, but without Beethoven we should never have had a C minor symphony. Alexander Moszkowski, *Einstein the Searcher* (London, 1921), p. 99. That is incontrovertible. Yet one may object, on the one hand, that the arbitrariness of scientific creation can be very well illustrated by Einstein's own example, the early history of the calculus; on the other hand, that the development of the symphony would have continued after Mozart, even if there had been no Beethoven. It remains true that Beethoven's work is absolutely individual and as such irreplaceable, while Newton's would have been replaced sooner or later by something equivalent.

at our right and at our left; this will help us to avoid them, if it be at all in our power to do so.

Bearing in mind these limitations, the study of the history of science will give scholars abundant fruits, and over and above that, it will give more wisdom to the wise, more human sympathy to those who are capable of it, and, to those whose souls are open to admiration rather than cynicism, new opportunities of admiring some of the greatest achievements of mankind.

BIBLIOGRAPHY

BIBLIOGRAPHY

This list, prepared for the beginner, will enable him, short as it is, to continue his bibliographical preparation to any extent. It is divided into seven sections:

I. Historical methods.

II. Scientific methods.

III. Chief reference books for the history of science. A. History and biography; B. Catalogues of scientific literature; C. Union lists of scientific periodicals; D. General scientific journals.

IV. Journals and serials on the history of science.

V. Treatises on the history of science.

VI. Handbooks on the history of science.

VII. Societies and congresses. A. History of science societies; B. National scientific societies; C. International congresses.

It should be noted that the history of each science has its own bibliography, and that all these bibliographies are somewhat overlapping. I propose to deal with the history of separate sciences on other occasions; meanwhile an intelligent student would have no insurmountable difficulties in compiling a special bibliography of the history of any science (or any subject) on the basis of the indications given below.

I. Historical Methods

The best known of treatises on historical methods are:

Ernst Bernheim (1850–), *Lehrbuch der historischen Methode* (Leipzig, 1889; often reprinted).

Charles Victor Langlois (1863–1929) and Charles Seignobos (1854–), *Introduction aux études historiques* (Paris, 1897; often reprinted).

I have never used Bernheim. The book of Langlois and Seignobos is excellent; it has been translated into English by G. G. Berry (London, 1898; reprinted in 1912, 1925, 1926).

Ch. V. Langlois, *Manuel de bibliographie historique* (Paris, 1901–04).

Louis John Paetow (1880–1928), *A Guide to the Study of Medieval History* (Berkeley, 1917). Revised edition prepared by the Mediaeval Academy of America (New York, 1931). A very useful book and not only for the mediaevalist.

By means of Langlois's books and Paetow's it is possible to explore the main historical literature, and become acquainted with historical methods in general.

For the general point of view of the historian of science, see G. Sarton, *The History of Science and the New Humanism* (New York, 1931; cf. *Isis*, vol. 16, pp. 451–455); Federigo Enriques, *Signification de l'histoire de la pensée scientifique* (Paris, 1934; 68 pp.; cf. *Isis*, vol. 23, p. 576); Abel Rey, *Les mathématiques en Grèce au milieu du V siècle* (Paris, 1935; 92 pp.; generalities at the beginning; cf. *Isis*, vol. 24, p. 470).

II. Scientific Methods

The only way to study scientific methods thoroughly is to work in a special field of science, and to carry on as many experiments and investigations as possible. Book knowledge cannot possibly replace the experimental knowledge obtained in the laboratory. Of course this is true also of historical methods, which can only be mastered by long practice.

However, for the historian of science, the experimental knowledge, indispensable as it is, is not sufficient. He must be more fully aware of the methods which scientists are applying to their purpose, and be able to analyze them.

There are a great many books dealing with the methods of science, and I could not tell which are the best, as I have read only a few. A good part of that subject is already standardized, and explained sufficiently well in every book.

Karl Pearson (1857–1936), *The Grammar of Science* (London, 1892; often reprinted). A pioneer work, still valuable.

Federigo Enriques, *Problemi della scienza* (Bologna, 1906). French translation (1909), German translation (1910), second Italian edition (1910), English translation entitled *Problems of Science* (Chicago, 1914; cf. *Isis*, vol. 3, p. 368). The author is a mathematician and the head of the institute for the history of science attached to the University of Rome.

Henri Poincaré (1854–1912), *La science et l'hypothèse* (Paris, 1908); *La valeur de la science* (1909); *Science et méthode* (1909). These books have been often reprinted and translated into many languages. The English translation of them by George Bruce Halsted with a special preface by the author and an introduction by Josiah Royce is available in a single volume (New York, 1913; many times reprinted).

Frederic William Westaway, *Scientific Method: Its Philosophical Basis and its Modes of Application* (London, 1912, later editions 1919, 1924, 1931; cf. *Isis*, vol. 4, pp. 119–122). On a much lower level than the preceding books, and thus more accessible to the average student. The author is an inspector of the English schools.

Arthur David Ritchie, *Scientific Method. An Inquiry into the Character and Validity of Natural Laws* (London, 1923). The author is a chemical physiologist.

André Lamouche, *La méthode générale des sciences pures et appliquées* (Paris, 1924). The author is an engineer in the French navy.

Abraham Wolf, *Essentials of Scientific Method* (London, 1925). Many times reprinted. The author is professor of the subject in the University of London, and he is also a historian of science.

Alfred North Whitehead, *Science and the Modern World* (Cambridge, 1926). The author is a mathematician and metaphysician.

Frederick Barry, *The Scientific Habit of Thought. An Informal Discussion of the Source and Character of Dependable Knowledge* (New York, 1927; cf. *Isis*, vol. 14, pp. 265–268). The author is a chemist, now professor of the history of science in Columbia University.

Norman Robert Campbell, *Physics, the Elements* (London, 1920); *An Account of the Principles of Measurement and Calculation* (London, 1928). The author is a physicist.

Harold T. Davis, *Philosophy and Modern Science* (Bloomington, Indiana, 1931; cf. *Isis*, vol. 18, pp. 204–206). The author is a mathematician and statistician.

Every student of the history of science should read at least Poincaré's immortal books, but that would not suffice. Poincaré's books are collections of essays, which do not replace more systematic treatises. With regard to the other books mentioned above, I will add two remarks.

1. Though these books have many elements in common, they are very different in contents and in spirit, and answer different purposes. Students need not read them all, but only a few; they should examine them and select those which seem the best for their own needs. They must be made to realize the concreteness of scientific methods and also their philosophical implications.

2. The list is exemplary rather than exhaustive. I have mentioned some of the books I have come across; there are many others with which I am not familiar and which may be as good, if not better.

III. Chief Reference Books for the History of Science

A. *History and Biography*

George Sarton, *Introduction to the History of Science*. Vol. 1. From Homer to Omar Khayyam (Baltimore, 1927). Vol. 2, in two parts. From Rabbi ben Ezra to Roger Bacon (1931). Vol. 3. Science and Learning in the Fourteenth Century (in preparation).

Ludwig Darmstaedter (1846–1927), *Handbuch zur Geschichte der Naturwissenschaften und der Technik*. Zweite Auflage (Berlin, 1908). Chronological list of discoveries year by year. Valuable, but to be used with caution.

August Hirsch (1817–94), *Biographisches Lexikon der hervorragenden Ärzte aller Zeiten und Völker* (1884–88, 6 vols.). New edition prepared by W. Haberling, F. Hübotter, and H. Vierordt (Berlin, 1929–34, 5 vols.; supplement, 1935).

I. Fischer, *Biographisches Lexikon der hervorragenden Ärzte der letzten fünfzig Jahren* (Berlin, 1932–33, 2 vols.).

These dictionaries of medical biography supplement Poggendorff's work, which is restricted to the exact sciences, and the more so because they contain biographies of many naturalists; indeed, down to the nineteenth century the majority of these were physicians.

Aksel G. S. Josephson, *A List of Books on the History of Science*. January, 1911. Supplement, December, 1916 (John Crerar Library, Chicago, 1911–17). Though this is only a library catalogue, and is already twenty years old, it is still a valuable tool, the authors' names and titles of books being quoted with great accuracy.

B. *Catalogues of Scientific Literature*

Johann Christian Poggendorff (1796–1877), *Biographisch literarisches Handwörterbuch zur Geschichte der exakten Wissenschaften* (Leipzig, 1863, 2 vols.). Vol. 3 for the period 1858–83 (1898). Vol. 4 for the period 1883–1904 (1904). Vol. 5 for the period 1904–22 (1926).

Royal Society of London, *Catalogue of Scientific Papers, 1800–1900* (Cambridge, 1867–1925, 19 vols.). Subject index (1908–14, 4 vols.).

This work is so important that we must pause a moment to describe it. Its compilation was first suggested at the Glasgow meeting of the B.A.A.S. in 1855 by Joseph Henry (1797–1878), secretary of the Smithsonian Institution, and the plan was drawn up in 1857. After many years of preparation and considerable expenditure, the first volume appeared in 1867, and the publication continued as follows:

First series. Vols. i–vi, cataloguing the papers of 1800–63, 1867–77.
Second series. Vols. vii–viii, literature of 1864–73, 1877–79.
Third series. Vols. ix–xi, literature of 1874–83, 1891–96.
Vol. xii. Supplement to the previous volumes, 1902.
Fourth series. Vols. xiii–xix, literature of 1884–1900, 1914–25.

To give an idea of the size of this catalogue it will suffice to remark that the papers catalogued in the fourth series alone, for the period 1884–1900, number 384,478, by 68,577 authors.

The compilation of a subject index, without which the work loses much of its value, was already contemplated in the first plan (1857). It was finally decided to arrange it in accordance with the *International Catalogue of Scientific Literature* (see below). This meant that it would include seventeen volumes, one for each of the seventeen sciences recognized in that catalogue. The first volume, Pure Mathematics, appeared in 1908; the second, Mechanics, in 1909, the third, Physics, in two instalments, Generalities, Heat, Light, Sound in 1912, Electricity and Magnetism in 1914. The publication seems to have been finally discontinued, which is a great pity. Whatever the fate of the International Catalogue may be, there is no justification for leaving the Royal Society Catalogue essentially incomplete, and thus nullifying a large part of the past labor and expenditure.

International Catalogue of Scientific Literature. Published for the International Council by the Royal Society of London.

This is an outgrowth of the Royal Society Catalogue, as it was felt that the scientific literature of our century was too extensive to be dealt with by a single scientific society. Its organization was arranged at the initiative of the Royal Society, by an international conference which met in London in 1896, then again in 1898, in 1900, etc. It was decided to divide science into seventeen branches:

A. Mathematics.
B. Mechanics.
C. Physics.
D. Chemistry.
E. Astronomy.
F. Meteorology (incl. Terrestrial magnetism).
G. Mineralogy (incl. Petrology and Crystallography).
H. Geology.
J. Geography (mathematical and physical).
K. Palaeontology.
L. General biology.
M. Botany.
N. Zoology.
O. Human anatomy.
P. Physical anthropology.
Q. Physiology (incl. experimental Psychology, Pharmacology, and experimental Pathology).
R. Bacteriology.

A large number of volumes were actually published from 1902 to 1916, but the gigantic undertaking was a victim of the World War and of the national selfishness and loss of idealism which the War induced. The volumes published cover the scientific literature for the period from 1901 to about 1913.[1]

[1] The publication includes 254 octavo volumes, varying in thickness from half an inch to two inches, and the original price was about £260. The stock has been sold to William Dawson and Sons, London, who offer a complete set for the price of £60 unbound, or £100 bound (November, 1935).

C. *Union Lists of Scientific Periodicals*

The two most important lists of that kind are:

1. The *Union List of Serials in Libraries of the United States and Canada* (New York, 1927, one very large quarto volume of 1588 pp.).

Registering some 70,000 journals and serials, of every kind, dead or alive, published in some 70 languages, and available in some 225 American libraries. Two supplements have already appeared, bringing the list down to 1932.

2. *A World List of Scientific Periodicals Published in the Years 1900–1933*. Second edition (London, 1934, large quarto, 794 pp.). Less comprehensive than 1 because it is restricted to contemporary scientific publications, it includes some 36,000 entries in 18 languages (for statistics, see *Isis*, vol. 23, p. 578).

These two lists are useful, first, to identify a certain journal, secondly, to find in what libraries (British or American) sets of it are available, and, finally, to judge of its importance, or at least of its popularity, by the number of sets available in the English-speaking world. This last judgment is possible only in the case of publications which are not distributed mostly by gift or exchange.

D. *General Scientific Journals*

For the study of science and the determination of the main impulses and tendencies of contemporary research, it is well to consult journals devoted to science in general. The ten leading journals of that kind are, in chronological order:

1. *Revue scientifique*. Paris, 1863.
2. *Nature*. London, 1869.
3. *La Nature*. Paris, 1873.
4. *Science*. New York, 1883.
5. *Naturwissenschaftliche Rundschau*. Braunschweig, 1886–1912.
6. *Revue générale des sciences pures et appliquées*. Paris, 1890.
7. *Science Progress in the Twentieth Century*. London, 1906.
8. *Scientia*. Bologna, 1907.
9. *Die Naturwissenschaften*. Berlin, 1913. (A continuation of 5.)
10. *Discovery*. London, 1920.

The richest in information of these journals is *Nature*. It contains by far the largest collection of material of this sort; by the end of 1935, 136 large volumes had been published. Each volume is indexed, but there are no general indices.

Of these journals the only ones having general indices are the *Revue scientifique* for the period 1863–81, *La Nature* for 1873–1912 (four decennial indices), the *Revue générale des sciences* for 1890–1914, and *Die Naturwissenschaften* for 1913–27. Thus through these four journals the general scientific achievements of the period 1863–1927 are indexed to some degree.

IV. Journals and Serials on the History of Science

The chief journals and serials are quoted in chronological order. The publishers named are in each case the latest or last.

1. *Klassiker der exakten Naturwissenschaften.* Founded by Wilhelm Ostwald (1853–1932). (Leipzig, 1889, etc.) Vols. 238, 239 appeared in 1934. Publisher, Akademische Verlagsgesellschaft, Leipzig.

2. *Mitteilungen zur Geschichte der Medizin und der Naturwissenschaften, herausgegeben von der Deutschen Gesellschaft für Geschichte der Medizin und der Naturwissenschaften* (vol. 1, 1902; in 1936, vol. 35 is being published). Almost exclusively bibliographical. Publisher, Leopold Voss, Leipzig.

3. *Archiv für die Geschichte der Naturwissenschaften und der Technik* (13 vols. published from 1909 to 1930; vol. 9 contains only 126 pp.). Publisher, F. C. W. Vogel, Leipzig.

4. *Rivista di storia critica delle scienze mediche e naturali* (vol. 1, 1910). Organ of the Italian Society for the History of Science. Direttore: Andrea Corsini, Via de' Bardi 5, Firenze (113).

5. *Isis.* Quarterly organ of the History of Science Society and bibliographical organ of the International Academy for the History of Science. Edited by George Sarton (vol. 1, 1913; the publication of vol. 26 began in 1936). Publisher, Saint Catherine Press, Bruges.

This is the chief journal devoted to the history of science and the most comprehensive. It includes new contributions, reviews, notes, abundant illustrations, and a very elaborate critical bibliography covering the whole field. That bibliography is arranged in the same order as Sarton's *Introduction*; it corrects and keeps up to date the volumes of the *Introduction* already published and accumulates materials in their proper sequence for the ulterior volumes.

6. *Studies in the History and Method of Science.* Edited by Charles Singer (1917–21, 2 vols. quarto, with illustrations). Published by the Clarendon Press, Oxford.

A splendid collection, unfortunately interrupted after the second volume.

7. *Archivio di storia della scienza*, later called *Archeion.* Edited by Aldo Mieli (vol. 1, 1919; in 1935, vol. 17 is being published; vol. 10 has never been completed). Official organ of the International Academy of the

History of Science, 12 rue Colbert, Paris II. Published by the Casa Editrice Leonardo da Vinci, Rome.

8. *Abhandlungen zur Geschichte der Naturwissenschaften und der Medizin.* Edited by Oskar Schulz (7 parts published, numbered 1 to 2 and 4 to 8. 1922–25). Publisher, Max Mencke, Erlangen, Bavaria.

9. *Quellen und Studien zur Geschichte der Mathematik, Astronomie, und Physik.* Edited by O. Neugebauer, J. Stenzel, and O. Toeplitz. Published in two sections, A, Quellen, and B, Studien. Section A began to appear in 1930 and section B in 1931.

10. *Quellen und Studien zur Geschichte der Naturwissenschaften und der Medizin.* Edited by P. Diepgen and J. Ruska. Continuation of the *Archiv* (no. 3). Began to appear in 1931. Nos. 9 and 10, complementing one another, are published by Julius Springer, Berlin.

11. *Archives for the History of Science and Technology* (in Russian). Published in Leningrad by the Academy of Sciences of the USSR (first volume, 1933, eighth volume, 1936). Regularly analyzed in *Isis* like all other periodicals, sometimes at greater length, as relatively few non-Russians are able to read Russian.

12. *Abhandlungen zur Geschichte der Medizin und Naturwissenschaften.* Edited by Paul Diepgen, Julius Ruska, and Julius Schuster (Heft 1, Berlin, 1934; Heft 8, 1935). Verlag Dr. Emil Ebering, Berlin.

13. *Thalès. Recueil annuel des travaux de l'Institut d'histoire des sciences et des techniques de l'Université de Paris.* Première année, 1934. Published by Alcan, Paris, in 1935. Edited by Abel Rey, Ducassé, Pierre Brunet.

14. *Osiris.* Studies on the History and Philosophy of Science, and on the History of Learning and Culture. Edited by George Sarton (vol. 1, Bruges, 1936).

This series is supplementary to *Isis*. It will include volumes devoted to a single subject or group of subjects (as vol. 1, devoted to the history of mathematics) and the longer and more technical papers; *Isis*, the shorter ones, the reviews, notes, queries, and critical bibliography.

15. *Annals of Science. A Quarterly Review for the History of Science since the Renaissance.* Edited by Douglas McKie, Harcourt Brown, and H. W. Robinson (London, 1936). Publishers, Taylor and Francis, London E.C. 4.

There are many more journals and serials devoted to the history of special sciences, especially the medical ones, but they can all be traced through the bibliographies above mentioned. In 1914, Sarton published a list of sixty-two reviews and collections devoted to the history of science (*Isis*, vol. 2, pp. 125–161); many other items could now easily be added to it through the critical bibliographies published in *Isis* ever since (no. 45 in vol. 25). Indeed, these bibliographies constitute a systematic and continuous table of contents to all the reviews and serials dealing with the history and philosophy of science, and to all the similar materials published in other journals or anywhere.[1]

Journals devoted to the history of science contain many articles on the philosophy of science, these two groups of subjects being closely related. Articles on the philosophy of science are also found in most philosophical journals. For example, the *Revue de métaphysique et de morale*, founded in 1893, by Xavier Léon (1868–1935) and edited by him (general tables for vols. 1 to 30, 1893–1923), has published a magnificent series of papers on the subject, including many of Poincaré's. A special journal, *Philosophy of Science*, edited by William Marias Malisoff, has been published by Williams and Wilkins in Baltimore since 1934.

V. Treatises on the History of Science

The earliest treatise deserving to be quoted here is William Whewell, *History of the Inductive Sciences from the Earliest to the Present Times* (London, 1837, 3 vols.). Vol. 1 deals with Greek physics, Greek astronomy, mediaeval physics, and astronomy from Copernicus to Kepler; vol. 2, with the history of mechanics, physical astronomy, acoustics, optics, "thermotics and atmology"; vol. 3, with other branches of physics, chemistry, natural history, physiology and comparative anatomy, and, finally, geology. Irrespective of its many shortcomings, some of which were unavoidable a century ago, this book is not a history of science as we understand it to-day, but a juxtaposition of various special histories, which is something very different; it represents a lower stage of integration.

Thus far there is only one modern complete treatise on the history of science. That is Friedrich Dannemann, *Die Naturwissenschaften in ihrer Entwicklung und in ihrem Zusammenhange* (Leipzig, 1910–13, 4 vols.; 2d ed. 1920–23; cf. *Isis*, vol. 2, pp. 218–222; vol. 4, pp. 110, 563; vol. 6, p. 115). It is elementary and imperfect, yet Dannemann, like Whewell, was a pioneer, and deserves our gratitude on that account.

When one remembers, on the one hand, the insufficiency of this treatise, and on the other hand, that it is not only the best but the only one of its kind, one realizes more keenly the immaturity of our studies and the immensity of the task to be accomplished.

Three new treatises have been begun but are only incompletely published:

Abel Rey, *La science dans l'antiquité* (Paris, 1930–33). This is a part of Henri Berr's collection *L'évolution de l'humanité*. Two volumes have appeared: vol. 1, La science orientale avant les Grecs; vol 2, La jeunesse de la science grecque (cf. *Isis*, vol. 21, pp. 224–226).

[1] As far as available to the editor of *Isis*.

Federigo Enriques and G. de Santillana, *Storia del pensiero scientifico*, vol. 1, Il mondo antico (Milano, 1932; cf. *Isis*, vol. 23, pp. 467–469).

Pierre Brunet and Aldo Mieli, *Histoire des sciences*. Antiquité (Paris, 1935; cf. *Isis*, vol. 24, pp. 444–447).

Of these three works the first is built on by far the largest scale, and is the most erudite. The two volumes thus far published by Rey carry the story only down to the middle of the fifth century B.C. Each of the two other works covers the whole of antiquity and the beginning of the Middle Ages in a single volume. Enriques's treatise is more philosophical; Brunet and Mieli's more elaborate. Moreover, the latter is a combination between a treatise and an anthology, about half of the space being sacrificed to a collection of extracts from the original writings in French translation. Further comparisons between these three works would be invidious.

To these books must be added one of a different kind, which is in a class by itself.

Lynn Thorndike, *A History of Magic and Experimental Science during the First Thirteen Centuries of our Era* (New York, 1923, 2 vols.; cf. *Isis*, vol. 6, pp. 74–89); and *in the Fourteenth and Fifteenth Centuries* (New York, 1934, 2 more vols.; cf. *Isis*, vol. 23, pp. 471–475).

These four volumes deal largely with the borderland between science and magic. They are very erudite, and historians of science should not fail to consult them.

It is interesting to note that the authors of all these treatises are professional students of the history of science, that is, men for whom that study has become the main activity. Even the old William Whewell (1794–1866), trained as a mathematician and physicist, devoted most of his time to the history and philosophy of science. Dannemann, now retired, was for many years a teacher of the history of science. Abel Rey, by training a philosopher, is professor of our subject at the Sorbonne and director of the Institute of the History of Science attached to the University of Paris; the mathematician Enriques is now director of a similar institute in the University of Rome and Santillana is his assistant; Mieli, who learned the methods of science as a chemist, is the founder and permanent secretary of the International Academy of the History of Science in Paris, and Brunet is a collaborator of his. Thorndike has devoted his whole life to the study of the interrelations between science and magic.

VI. Handbooks on the History of Science

When scholars are beginning to take an interest in our studies, their first query is, naturally enough, "Could you recommend a single volume giving an outline of the whole subject?" Such a volume does not yet exist, and this is not surprising when one knows how the matter stands with regard to treatises. Elementary books can only be written in a satisfactory way when elaborate treatises are available. It is possible to-day to write a little book covering the whole of, say English literature, or the Reformation, or any other standardized subject, and to be confident that, however small the scale, nothing essential, from the standpoint of that scale, is likely to be overlooked. For the history of science such a feat of selection and compression is still impossible, because the introductory analyses and surveys have not yet been completed; or, if not impossible, it is very much of a wager and a gamble.

A few years ago the veteran historian of science Siegmund Günther wrote a little book, *Geschichte der Naturwissenschaften* (Reclam series, 1909). The degree of selection being much too small for the degree of compression, the book is unreadable: it is more like a catalogue than a story. It is as if one tried to crowd too many names on a small scale map.

The best single volume available to-day is Sir William Cecil Dampier Dampier-Whetham's *A History of Science and its Relations with Philosophy and Religion* (Cambridge, 1929; cf. *Isis*, vol. 14, pp. 263–265).

William Thompson Sedgwick (1855–1921) and Harry Walter Tyler, *A Short History of Science* (New York, 1917). A primer, which I only quote for lack of better. The surviving author, Tyler, is preparing a new edition, which will certainly be an improvement.

Charles Singer has been working on an introductory volume for many years, the *Short History of Science*, to be published soon by the Clarendon Press. Another book almost ready for publication is Benjamin Ginzburg's *Origins of Modern Science.*

If we had to select a guidebook to Europe, purporting to indicate and to explain within the covers of a single volume the chief curiosities of the whole continent, our first question would concern the personality of the author. Of course we should have more confidence in him if we knew he had himself travelled all over Europe than if we discovered that he had compiled his guide in the New York Public Library. In a similar way, for the appreciation of a handbook on the history of science, the prime consideration must be the wisdom and experience of the writer.

Siegmund Günther (1848–1923) was one of the founders of the history of science movement in Germany, and the author of many books and memoirs on the history of the mathematical and physical sciences. Sir William Cecil Dampier Dampier-Whetham is a physico-chemist, but for the last twenty-five years he has devoted much time and thought to the history and the cultural aspects of science. Sedgwick and Tyler, the first a biologist, the second a mathematician, taught the history of science for many years at the Massachusetts Institute of Technology. Charles Singer, a physician, is the leading historian of medicine and biology in the British Empire.

VII. Societies and Congresses

A. *History of Science Societies*

Academies of science and scientific societies ocasionally take an interest in the history and philosophy of science. For example, the Académie des Sciences of Paris awards every year (since 1903) a prize, the Prix Binoux, to encourage work in that field (*Isis*, vol. 8, p. 161; vol. 25, p. 136). Moreover, the older academies and societies are naturally concerned with their own glorious past, with the history of their achievements and institutions and the biographies of their members, and this has often induced them to promote historical investigations.[1] The jubilee publications of those bodies sometimes contain historical memoirs of real value, which do not always receive the publicity they deserve and thus are relatively unknown.

However, the attention paid to our studies by those societies whose main interests are different from ours is erratic, and not much assistance can be expected from them. Therefore it has been found necessary to create special societies devoted to the study of the history of science.

The five principal ones are, in chronological order:

1. 1901. Deutsche Gesellschaft für Geschichte der Medizin und der Naturwissenschaften. Founded at Hamburg, September 25, 1901, by Karl Sudhoff and others. Publishes the *Mitteilungen zur Geschichte der Medizin und der Naturwissenschaften*, 1902 ff.

Annual dues, RM.20. Address: Rudolf Blanckertz, Georgenkirchstr. 44, Berlin NO 43, Germany.

2. 1907. Società italiana di storia critica delle scienze mediche e naturali. Founded at Perugia, October 9, 1907, by Domenico Barduzzi

[1] Two academic undertakings deserve special mention because of their unusual amplitude and quality:

1. The reports prepared by the Institut de France by order of Napoleon on the progress of science from 1789 to 1810. J. B. J. Delambre, *Rapport historique sur les progrès des sciences mathématiques depuis 1789 et sur leur état actuel* (272 pp.). Including mechanics, astronomy, geography, arts and industries. Georges Cuvier, *Rapport historique sur les progrès des sciences naturelles* (298 pp.). Including chemistry, physics, physiology, natural history, medicine, agriculture. André Dacier, *Rapport historique sur les progrès de l'histoire et de la littérature ancienne* (263 pp.). The three quarto volumes were published at Paris in 1810.

2. The series of histories of particular sciences prepared by order of the Royal Academy of Sciences of Bavaria. A series of some 24 volumes, which began to appear about 1864 and was completed about half a century later. It is the finest series of its kind.

(1847–1929) and others. Publishes *Atti delle riunioni*, 1909 ff., and *Rivista di storia critica delle scienze mediche e naturali*, 1913 ff.

Annual dues: L.25 in Italy, L.50 abroad. Address: Prof. Andrea Corsini, Via de' Bardi 5, Firenze (113), Italy.

3. 1924. History of Science Society. Founded in Boston, January 12, 1924, by David Eugene Smith and others. Publishes *Isis* and various books; the members receive *Isis* free of charge, but not the other publications.

Annual dues: $5. Address: Frederick E. Brasch, Library of Congress, Washington, D. C., U. S. A.

4. 1928. Académie internationale d'histoire des sciences. Founded in Oslo, August 18, 1928, by Aldo Mieli and others. Its official organ is *Archeion*, and its bibliographic organ, *Isis*.

Membership restricted to a hundred elected members. Address: Perpetual secretary, Aldo Mieli, 12 rue Colbert, Paris 2, France.

5. 1935. Lärdomshistoriska Samfundet (Society of the History of Learning). Founded in Upsala in 1934 by Johan Nordström and others. Publishes Lychnos, *Lärdomshistoriska Samfundets Årsbok*, an annual review; vol. 1, 1936.

Annual dues: 8 crowns. Address: Prof. J. Nordström, University, Upsala, Sweden.

This society, the youngest of all, is also by far the largest in membership. It already counted 1750 members in May, 1935, before its publications began to appear.

Each of these societies and of others less important or devoted to a single science or group of sciences (as medicine) is a centre of research for our studies. In addition to their regular and irregular publications, they organize annual meeetings, lectures, and discussions.

B. *National Scientific Societies*

Many countries have organized annual scientific congresses, the importance of which cannot be overestimated. It is true, they do not much influence the progress of science, which is taken care of more effectively by the academies and the special scientific societies, but they are very powerful in diffusing scientific knowledge and the appreciation of scientific methods, and in moulding public opinion. The parent and model of these national associations is the Gesellschaft deutscher Naturforscher und Ärzte, which met for the first time at Leipzig in 1822, and has met every year since, each time in a different town of Germany or Austria. The other national societies also move from town to town and thus establish each year a new centre of diffusion. The proceedings of these societies enable historians of science to determine the main trends in science year by year; moreover sections of those societies are or may be devoted to our studies.

The five main national societies are, in chronological order:

1. 1822. G.d.N.A. Gesellschaft deutscher Naturforscher und Ärzte. Unfortunately the publications are very irregular. The reports (*Verhandlungen* and other titles) of the early meetings appeared in the *Isis* of Lorenz Oken (1779–1851) who was the founder of the Gesellschaft; in 1836, the *Tageblatt der Versammlung* began to appear; since 1924 the *Mitteilungen* are published as a supplement to *Die Naturwissenschaften*. I know of no general guide or indices to all these proceedings, nor have I ever seen a complete set of them.

Offices: Gustav Adolfstr. 12, Leipzig C 1.

A section is devoted to the history of medicine and natural sciences, its meeting being organized in conjunction with the German society *ad hoc*. The proceedings of that section are generally published in the third or Dutch *Janus*,[1] one of the leading journals on the history of medicine founded in 1896 and published at Haarlem and later at Leyden.

2. 1831. B.A.A.S. British Association for the Advancement of Science. This association met for the first time at York in 1831, and has met every year since in a different town of Great Britain or the British Empire. Its influence upon the English speaking peoples has been quite considerable. Its *Reports* published annually in separate volumes since 1831, with general indices for the years 1831–60 and 1861–90, constitute a valuable collection for the historian of science (as opposed to the German reports, which being scattered and irregularly published are so difficult to consult in their entirety that one does not try to).

No special section of the B.A.A.S. is devoted to the history of science. Address: Burlington House, London, W. 1. The official residence of the Permanent Secretary is now at Down House, at Downe, Kent, formerly Darwin's home (see *Isis*, vol. 23, pp. 533, 534).

3. 1848. A.A.A.S. American Association for the Advancement of Science. *Proceedings* published in annual volumes until 1910. Since then the full proceedings appear in *Science*, and only Summarized Proceedings from time to time in book form. No indices. Section L is devoted to the 'historical and philological sciences.' The annual meetings of the History of Science Society are arranged to coincide every other year with the meetings of section L, and on the alternate years with the meetings of the American Historical Association.

Address of A.A.A.S.: Smithsonian Institution, Washington, D. C.

4. 1872. A.F.A.S. Association Française pour l'Avancement des Sciences. *Comptes-rendus*, published every year, no indices. There is no

[1] Two other medico-historical journals bearing the same name preceded the Dutch *Janus* (see *Isis*, vol. 2, p. 143, 146; vol. 17, pp. 283–284).

section devoted to the history of science as such, but there is an archaeological section.

Address: Secrétariat de l'Association, rue Serpente 28, Paris VI.

5. 1907. S.I.P.S. Società Italiana per il Progresso delle Scienze. *Atti*, published each year.[1] Elaborate decennial indices for 1907–19, 1921–31. Section 1 of class C deals with history and archaeology, section 2 with philology, section 5 with philosophy. Address: Via del Collegio Romano 26, Roma.

C. *International Congresses*

International congresses have been organized for almost every science or group of sciences, and one of their sections is generally devoted to historical subjects. It is not possible to deal here with these congresses, any more than with the national societies relative to special sciences. In my book on the study of the history of mathematics, I shall speak of the mathematical societies and congresses. However, this is the proper place to speak of two other series of international congresses which concern the history and philosophy of science in general; these are the congresses of history and philosophy.

In addition to two international meetings — at Chicago, 1893, and The Hague, 1898 — which are not counted in the regular series, the international congresses of historical sciences have taken place as follows:

I. Paris, 1900.
II. Rome, 1903.
III. Berlin, 1908.
IV. London, 1913.
V. Brussels, 1923.
VI. Oslo, 1928.
VII. Warsaw, 1933.

The international congresses of philosophy have met as follows:

I. Paris, 1900.
II. Geneva, 1904.
III. Heidelberg, 1908.
IV. Bologna, 1911.
V. Naples, 1924.
VI. Cambridge, Mass., 1926.
VII. Oxford, 1930.
VIII. Prague, 1934.

The proceedings of these congresses, especially of the philosophical ones, contain valuable materials for the historians of science. For example, Paul Tannery took a leading part in the historical congress of

[1] No meetings took place in 1914, 1915, 1918, 1920, 1922.

Rome (1903) and in the philosophical congresses of Paris (1900) and Geneva (1904). The special section under his chairmanship of the Geneva meeting was perhaps the most brilliant congress of the history of science which ever occurred.

An international congress of scientific philosophy, the first of its kind,[1] took place in Paris, September, 1935; a second meeting will take place in Copenhagen 1936.

To conclude, mention must be made of the International Congresses of the History of Science organized by the Academy *ad hoc*. Three congresses have thus far taken place:

I. Paris, 1929.
II. London, 1931.
III. Coimbra, 1934.

and the following are in preparation:

IV. Prague, 1937.
V. Lausanne, 1940.

No proceedings have been published of the first two congresses, except for summaries which appeared in *Archeion* and *Isis*, and the publication of individual papers in various journals. For the third one, see Actes, conférences et communications du IIIe Congrès international d'histoire des sciences (Lisbon 1935). It is expected that these congresses will become gradually more and more important, because the problems interesting historians of science are generally too technical to interest conventional historians, and too concrete and limited to interest conventional philosophers.

[1] Yet every congress of philosophy has been also, through one of its sections, a congress of scientific philosophy. Philosophers are generally far more interested in the philosophy than in the history of science, and there seems to be less justification for the organization of special congresses for the philosophy than for the history of science.

INDEX

INDEX

I. PERSONS

II. TITLES OF JOURNALS, SERIALS, AND INSTITUTIONS